Gene Transfer and the Ethics of First-in-Human Research

Gene Transfer and the Ethics of First-in-Human Research

Lost in Translation

Jonathan Kimmelman
McGill University / Biomedical Ethics Unit and Social Studies of Medicine

CAMBRIDGE UNIVERSITY PRESS
Cambridge, New York, Melbourne, Madrid, Cape Town, Singapore, São Paulo, Delhi, Dubai, Tokyo

Cambridge University Press
The Edinburgh Building, Cambridge CB2 8RU, UK

Published in the United States of America by
Cambridge University Press, New York

www.cambridge.org
Information on this title: www.cambridge.org/9780521690843

First published 2010

Printed in the United Kingdom at the University Press, Cambridge

A catalogue record for this publication is available from the British Library

Library of Congress Cataloguing in Publication data
Kimmelman, Jonathan.
Gene transfer and the ethics of first-in-human research : lost in translation /
Jonathan Kimmelman.
p. cm.
Includes bibliographical references and index.
ISBN 978-0-521-69084-3 (pbk.) 1. Gene therapy–Moral and ethical aspects.
2. Clinical trials–Moral and ethical aspects. I. Title.
[DNLM: 1. Gene Transfer Techniques–ethics. 2. Human Experimentation–ethics.
3. Public Policy. WB 60 K49g 2010]
RB155.8.K56 2010
615.8′95–dc22 2009029830

ISBN 978-0-521-69084-3 Paperback

For Chichester and Gooch

We grow accustomed to the Dark --
When light is put away --
As when the Neighbor holds the Lamp
To witness her Goodbye --

A Moment – We uncertain step
For newness of the night --
Then – fit our Vision to the Dark --
And meet the Road – erect --

And so of larger – Darkness --
Those Evenings of the Brain --
When not a Moon disclose a sign --
Or Star – come out – within --

The Bravest – grope a little --
And sometimes hit a Tree
Directly in the Forehead --
But as they learn to see --

Either the Darkness alters --
Or something in the sight
Adjusts itself to Midnight --
And Life steps almost straight.

Emily Dickinson

Contents

Acknowledgments

This book began inauspiciously enough when, newly arrived in Montreal with a very bad haircut and a Ph.D. in molecular genetics, I approached Kathy Glass about post-doctoral studies in bioethics. I am grateful to Kathy for taking a risk by hiring me (first as a post-doc, and later as an assistant professor) and supporting this research.

I also thank Nancy King for her inspiring work in research ethics, as well as her support and critical feedback at key junctures. My colleague Nick King (no relation) provided a fine-grained vetting of this book. Alex London was a ready and more than able intellectual sparring partner throughout the writing process. Rebecca Dresser and Eric Juengst were gratuitously generous for agreeing to read my manuscript.

Numerous other oversubscribed colleagues, friends, and strangers made important contributions. These include: James Anderson, Françoise Baylis, Alberto Cambrosio, Marina Emborg, Dean Fergusson, Abe Fuks, Margaret Lock, Josephine Nalbantoglu, Tim Ramsey, Jason Scott Roberts, Thomas Schlich, Stan Shapiro, Charles Weijer, George Weisz, and Allan Young. I thank the many gene-transfer researchers, clinical investigators, and laboratory scientists (too numerous to name – and many probably preferring anonymity anyway) who graciously and patiently entertained my questions over the years. I also thank my editors at Cambridge, Katie James, Nick Dunton, Matt Davies and John Forder.

I received feedback from many anonymous commentators through peer review of various articles relating to this book. Some of these individuals should be shot (you know who you are). But many others volunteered elaborate and thoughtful criticisms that went well beyond the call of duty. Their anonymous service is an inspiration.

Through its main health granting agency, Canada "been very very good to me." I thank the Canadian Institutes of Health Research for supporting this work with various grants and awards, including the Maud Menten New Investigator Prize, a CIHR New Investigator Award (Ethics in Translational Research: Navigating the Interface between the Bench and Clinic, MSH 87725), and several operating grants (States of Mind: Emerging Issues in Neuroethics, NNF 80045; Therapeutic Hopes and Ethical Concerns: Clinical Research in the Neurosciences, MOP 77670; Investigators as Inventors: An Empirical Study of Patent Holding and Ethics in Human Gene Transfer Research, MOP 79288; Advancing the Ethics of Clinical Trials: Enhancing Participant Protection and Scientific Rigour and a New Investigator Award, MOP 68835). This funding has allowed me to employ two resourceful and reliable research assistants over the years, David McLaughlin and Kat Duckworth.

This book emphasizes the role of social and technical networks for managing uncertainty and risk. What is true for novel medical interventions also applies for "risky" book projects like this. My cup runneth over with social infrastructure. First thanks go to my parents, Harold and Joyce Kimmelman, for enthusiasm and support. Harold also helped with the graphics and cover design. My wife, Sara Laimon, has been tireless in helping to maintain and restore order and morale throughout the book writing. She further offered very useful intellectual and stylistic feedback. I also thank several close friends for their care and feedback on various issues. They include: Cristiano Martello, Sean Hecht, Todd Krichmar, Naomi Schrag, and Laura Wolf-Powers. Finally, my daughters Rachel and Tamara had small hands in this project; I thank them for sustaining me with their humor, humanity, and reprieve.

Every effort has been made to ensure the accuracy of this book's content. None of the above acknowledgments should be interpreted as suggesting endorsement, sanction, confirmation, collusion, blandishment, complicity, or financial interest in this work or its author. Despite the use of high standards and care in writing this book, certain errors of analysis might have occurred. Money for harm or wages lost as a result of faulty analysis is not available. By reading this statement, you understand that I take full responsibility for any faults, foreseeable or unforeseeable, in the analysis.

Abbreviations

AAV	adeno-associated virus
AAMC	Association of American Medical Colleges
ADA	adenosine deaminase
AHCJ	Association of Health Care Journalists
ASGT	American Society of Gene Therapy
CIOMS	Council for International Organizations of Medical Sciences
CNS	Central Nervous System
CRADA	cooperative research and development agreement
CTEP	cancer therapy evaluation program
CTSC	Clinical and Translational Science Center
DARPA	Defense Advanced Research Project Agency
DNA	deoxyribonucleic acid
DSMB	Data Safety Monitoring Board
ESGT	European Society of Gene Therapy
FDA	Food and Drug Administration
FH	familial hypercholesterolemia
GAD	glutamic acid decarboxylase
GIPAP	Glivec International Patient Assistance Program
hESC	human embryonic stem cell
HGTS	Human Gene Therapy Subcommittee
HHS	Health and Human Services
HIC	high-income country
HIV	human immunodeficiency virus
IBC	Institutional Biosafety Committee
IHGT	Institute for Human Gene Therapy
IND	investigational new drug
IRB	institutional review board
LCA	Leber's congenital amaurosis
LMIC	low-and middle-income country
MPS-VII	mucopolysaccharidosis type VII
MPTP	1-methyl-4-phenyl-1,2,3,6-tetrahydropyridine
NCI	National Cancer Institute
NGVB	National Gene Vector Biorepository
NGVL	National Gene Vector Laboratories
NHF	National Hemophilia Foundation
NHLBI	National Heart Lung and Blood Institute
NIH	National Institutes of Health
NUCDF	National Urea Cycle Disorders Foundation
OBA	Office of Biotechnology Activities
OTCD	ornithine transcarbamylase deficiency

PI	principal investigator
RAC	Recombinant DNA Advisory Committee
RNA	ribonucleic acid
RNAi	inhibitory ribonucleic acid
RT-PCR	reverse transcriptase polymerase chain reaction
SCID	severe combined immunodeficiency
siRNA	small interfering RNA
STAIR	Stroke Therapy Academic Industry Roundtable

Chapter 1

Introduction: gene transfer lost in translation

Introduction

French and British researchers have treated X-linked severe combined immune deficiency syndrome (X-SCID), otherwise known as "bubble boy" disease. Italian and UK researchers have also treated a related disease, ADA-SCID. Though only three patients were enrolled in the study, Swiss and German researchers have treated yet another severe immune disorder, chronic granulomatous disease. And some commentators believe American researchers are on the cusp of a durable treatment for hemophilia B. In these instances, it would appear that gene transfer – briefly, the administration of genetic materials to human beings – has finally earned the title of gene *therapy*.

But the field's development has been, and continues to be, a long, strange trip. As I write, the most visible name associated with gene transfer, W. French Anderson, is serving a fourteen-year prison sentence on child molestation charges.[1] Another leading figure, James Wilson, has nearly finished a five-year, FDA-imposed ban on leading clinical studies.[2] Other sanctions in the field's thirty-year history include one of the earliest ever violations of rules for human research issued by the US Department of Health and Human Services (Martin Cline, for initiating a study without proper IRB review[3]), and a widely publicized rebuke of two other leading figures (Ronald Crystal and Jeffrey Isner) for not reporting trial deaths to the NIH.[4] In 1995, a high-level panel at the NIH faulted the field for rushing into clinical trials.[5] In 2000, two prominent researchers editorialized in the pages of *Science* magazine "gene therapy has many of the worst examples of clinical research that exist."[6]

Doubts surround even gene transfer's triumphs. Five children cured in the X-SCID studies, for example, developed a rare leukemia linked to the gene transfer; one subsequently died. In the chronic granulomatous disease study mentioned above, genetically modified tissues expanded in a manner such that some researchers worry that it might signal the start of a cancer.[7] Some enthusiasts in the field like to cite the fact that regulatory authorities in China have licensed Gendicine®, the first ever gene transfer treatment to be commercialized anywhere. Skeptics, however, question the quality of evidence supporting licensure, inconsistencies between Chinese- and English-language published reports, and conflicts of interest in the regulatory process.[8,9]

Clearly, the process of translating the idea of gene transfer into clinical application has proven far more difficult than anticipated. Some of the difficulties are the product of untoward vectors (which shuttle genetic materials into cells) and immune systems (which seem bent on attacking vectors and the cells that receive them). Other difficulties include the untoward conduct of clinical researchers and investors. What I would like to advance in this book is the proposition that some of gene transfer's difficulties have arisen from a failure to translate ethical concepts for medical research, which were developed largely for controlled clinical trials (where compounds are relatively simple, and uncertainties bounded), to the context of small-scale, scientifically intensive translational trials (where uncertainty is unbounded). What I mean to suggest is not that gene-transfer researchers have been delinquent in their adherence to clinical research ethics, though there certainly have been instances of this. Rather, all parties to the research – investigators, ethicists, IRB members, policy makers, patients, and research advocates – have appeared lost in trying to apply ethical concepts that are not well matched to early-phase tests of novel interventions.

In 1997, Leroy Walters and Julie Gage Palmer echoed a sentiment expressed as early as 1983 in a US Presidential Commission report:[10] that somatic gene transfer (that is, genetic modification of tissues that will not be passed on to future generations) could be considered, from an ethical standpoint at least, a "natural and logical extension of the current techniques for treating disease."[11] This position has gone largely unchallenged. Yet in emphasizing the continuity of gene transfer with conventional medicine, I believe the "gene transfer as natural extension" position obscures the distinctive – if not quite unique – ethical and social challenges associated with translating gene transfer into clinical application.

Hemophilia and gene transfer: some challenges

Recent gene transfer trials for hemophilia B illustrate my argument. In several respects, clotting disorders like hemophilia represent the perfect systems for validating basic principles of gene transfer. Whereas many genetic diseases require delivering genes to specific parts of the body, coagulant factors (proteins involved in triggering clotting) are secreted and can thus be expressed from easily accessed tissues like muscle. Second, hemophilia therapies have a wide therapeutic window. That is, coagulant factor levels as low as 5 percent of normal amounts are sufficient for converting severe hemophilia into a mild form; investigators need only to achieve a modest elevation in factor levels in order to achieve clinical correction.

From a scientific perspective, hemophilia is appealing because factor protein circulates in the bloodstream, which means that investigators can monitor gene expression with a simple blood test (other diseases require complicated imaging or invasive biopsies). Additionally, factor production can be quantified objectively and reproducibly by assaying plasma factor concentration. Finally, hemophilia offers various ethical and safety advantages over other disease models. Few areas of translational research have animal models as effective and faithful as those used in hemophilia. Subjects, because they have access to recombinant factor replacement therapy, are less likely to be impelled to trial participation by medical desperation. And having survived blood contamination scandals, the hemophilia community is politically energized, scientifically sophisticated, and more likely to view invitations to participate in clinical research with an appropriate level of caution.

As with any clinical research protocol, a proposal to conduct a human gene transfer study in hemophilia patients would be evaluated on the basis of its adherence to a set of widely agreed-upon ethical principles: respect for persons, beneficence, and justice.[12] As to how to implement these principles, ethicists and policy makers have evolved a set of frameworks and practices. Many of these map awkwardly, if at all, to the types of small-scale, scientifically intense studies that characterize gene-transfer trials.[13] Consider the following:

- *When is the appropriate time to initiate clinical testing?* Discussions around various hemophilia trials show this to be a major point of contention. The standard, controlled clinical trial answer would draw on the concept of clinical equipoise – that is, uncertainty among the expert clinical community as to the relative advantage of either arm in a comparative trial. The concept of clinical equipoise was originally intended to guide investigators conducting controlled clinical trials. Translational studies, however, are usually uncontrolled (that is, they do not involve other study arms that enable comparison with another intervention). There is, as yet, no ethical "indicator" of a translational trial's ripeness.

- *How should the validity and value of a translational trial be judged?* Standard accounts would suggest the ability to perturb clinical equipoise for randomized, controlled trials or, in the case of phase 1 studies, the ability to provide sufficient information for phase 2 testing. In fact, very few gene-transfer trials lead directly to phase 2 testing. For example, five phase 1 studies have been conducted for hemophilia gene transfer to date; none have produced a phase 2 trial. Does this mean that the studies are valueless, or does it mean that the way many investigators, subjects, and ethicists conceive the value for translational studies is somehow flawed?
- *What is fair subject selection?* One little-noticed phenomenon in gene-transfer research is the recruitment of subjects from low- and middle-income countries. One of the most significant hemophilia gene-transfer trials conducted to date, for example, recruited subjects from Brazil (successfully) and India (unsuccessfully). Standard accounts of justice would require providing a plan for post-trial access to medications developed in a study, and an assessment of a study's responsiveness to health needs in the host community. Both have clear implications for large, late-stage trials involving prevalent disorders that are exported to low-income countries. How do the ethics change when studies are *importing* subjects to participate in studies of rare diseases?
- *What is the appropriate population in which to test a study drug?* In later-phase research, this question is settled by an assessment of a study's risks and direct medical benefits. Toxic oncology drugs, thus, are tested in cancer patients who have a chance of benefiting from the study drug (or, at least, less to lose should the drug prove toxic). In translational research, however, diseases that present the best opportunity for gathering knowledge often offer a less favorable balance of risks and direct benefits. Persons with hemophilia, for example, are medically stable – at least with respect to their clotting deficiency – because validated therapies are already available. When hemophilia patients enter early-phase trials of novel agents, they place their stable health status at risk. When, and under what conditions, is it ethically defensible to recruit medically stable subjects into a study testing a novel intervention with indeterminate risks?
- *How should study risks be evaluated, and how should they affect study design?* Unlike later-stage studies, first-in-human trials often have little or no human experience on which to draw in performing a risk–benefit analysis. Risk assessment, as such, involves much intuition, and surprises occur with unsurprising frequency. In various hemophilia gene-transfer trials, these have included stronger than anticipated immune responses to vectors, higher than expected expression of the corrective gene, and vector contamination of subjects' semen. How should we define a "good guess" of study safety? And how might protocols be designed to minimize or manage the unusually high levels of indeterminacy?
- *How should risks to study volunteers be appraised?* Research ethics has evolved a well-developed framework for ethically evaluating risk in human studies. Under this framework, known as component analysis, research risks are divided into those deriving from procedures that have a therapeutic warrant (e.g. study drugs) and risks deriving from those that do not (e.g. extra medical tests). The demarcation of therapeutic and non-therapeutic procedures is generally straightforward for randomized controlled trials. But for first-in-human phase 1 studies, the demarcation is far from obvious. Should

study drugs be considered to have therapeutic warrant even for diseases like hemophilia, where an effective standard of care is well established? Even if, for safety reasons, initial volunteers are to receive doses below which investigators expect to observe a therapeutic response? Or should vector risks be classified as research risks, despite the fact that animal evidence suggests that volunteers might benefit medically from participation?

- *How should risks to others be evaluated?* Most drug trials involve only minor burdens for persons other than the trial subject. Gene-transfer studies, however, occasionally pose ethically important risks to non-subjects. In one of the most successful hemophilia trials to date, gene transfer vector sequences were detected in the semen of six out of seven subjects, raising concerns about the possibility of transmission to sexual partners or offspring. In this case, it turned out that the vector sequences only transiently contaminated semen, and appeared not to have actually modified subjects' sperm. Nevertheless, the episode raises important questions about incorporating inter-generational risks in the ethical appraisal of some trials.

Gene transfer as a model for translational research

Over the past several years, medical centers have announced major gifts directed toward translational research.[14,15] Universities have broken ground on new translational clinical research facilities,[16] and the NIH recently announced 12 grants of unprecedented size to create infrastructure for translational trials.[17] The grants are part of the NIH's "Roadmap", a program to accelerate the application of basic biomedical discoveries.[18] The United Kingdom has a counterpart: a network of eight centers for early-phase oncology studies.[19] Perceiving an inordinate lag between discovery and drug development, the Food and Drug Administration recently issued new guidelines to stimulate first-in-human studies.[20] The European Organization for Research and Treatment of Cancer announced new efforts to coordinate translational research efforts and strengthen their scientific basis.[21]

If these and other initiatives are to succeed – and avoid some of the challenges that have befallen gene transfer – it seems critical that an ethical framework particular to translational research be elaborated. This book offers a series of proposals in the hopes first of improving current practice, and second, stimulating further ethical analysis.

Gene transfer offers a particularly promising "model organism" in which to observe the emergence of ethical problems in translational research. First, the past two decades have witnessed extraordinary investment in genetic and genomic research. Gene transfer represents perhaps the culmination of these investments, and will likely play a significant role in enabling the application of other treatment strategies like cell and immunological therapies. Second, perhaps because of its medical promise – and its associated controversies – gene transfer has received an unusual degree of public scrutiny. News reports of advances and setbacks are therefore abundant. Third, gene-transfer research is more transparent than other areas of translational research. The operations of the NIH's Office of Biotechnology and its Recombinant DNA Advisory Committee afford the ethicist a rich record of ethical deliberation and practices. Finally, the pattern and organization of gene transfer research are characterized by an unusual degree of cohesiveness and collaboration. For example, professional societies like the American Society of Gene Therapy and the European Society of Gene Therapy have sprung up, as have medical journals. As will be discussed in the next chapter, gene transfer exemplifies a trend toward large, dispersed, collaborative, transdisciplinary networks of basic researchers, clinicians, and investors in translational research.

In portraying gene transfer as the translational research ethicist's *Drosophila*, I do not want to deny important discontinuities with other translational research realms. For example, gene-transfer agents are far more complex than targeted, small-molecule drugs or monoclonal antibodies. As well, the field has historically centered on low-incidence genetic diseases (though the majority of gene transfer protocols involve non-hereditary cancers and cardiovascular diseases). These don't significantly undermine my thesis, however. Human beings don't have wings, bristles, and compound eyes; model systems always have their limits.

Plan for the book

Gene transfer is complex, and many of the ethical issues encountered in this book articulate with complicated scientific questions. Chapter 2 introduces the science, sociology, history, and ethics of gene transfer, and highlights some of the features about gene transfer that make it distinctive from ethical, social, and policy standpoints.

In Chapters 3 and 4 I examine gene transfer's most conspicuous ethical challenge: risk. I use the death of a volunteer in a 1999 experiment, and the lymphoproliferative disorders observed in five volunteers participating in an X-SCID study, as springboards for addressing what makes gene-transfer research risky, how risks might be evaluated, and how uncertainty should be managed in novel research arenas.

Chapter 5 turns to the question of medical benefits for study volunteers. Is a translational trial a form of therapy, or is it a vehicle for delivering care to desperate patients? This chapter looks askance at claims that early-phase trials of novel interventions have therapeutic value. But it also questions the drawing of too sharp a distinction between care and therapy.

Another point of contestation that has received very little attention is the purpose of a phase 1 trial. Arrayed around this question are clinicians and sponsors, who view the first-in-human studies as a pivotal step in designing phase 2 trials, and bench scientists who take a longer view in seeing phase 1 studies as a way of pursuing scientific questions. Chapter 6 offers an analysis of value and validity as they pertain to translational research trials, and explores the implications of this analysis for study design, assessment of risk, and informed consent.

Chapter 7 turns to the question of when to initiate human trials. Astonishingly, research ethics has yet to provide a sustained analysis of the ethical basis for launching first-in-human studies. This chapter presents a framework – the principle of modest translational distance – that might provide some ethical guidance for deciding trial initiation. It also suggests factors that investigators, IRBs, policy makers, ethicists, and others should consider in deciding whether a research program has sufficient maturity to move into human subjects.

Chapter 8 examines questions of justice and fairness in translational research. As is true for later-stage research, many early-phase gene-transfer trials recruit volunteers from low and middle-income countries. However, the prevailing ethical framework for deciding whether such studies fulfill fair subject selection has limited applicability to early-phase gene-transfer studies. This chapter explores various promising avenues that researchers might explore in order to comply with international consensus on transnational research.

Many commentators have castigated the field of gene transfer for "overselling" the technology's promise and "spinning" trial results. In Chapter 9 I reframe ethical concerns underlying these criticisms as problems of expectation management, and explore how various investigators and leadership figures have attempted to shape the public reception of both favorable and unfavorable developments. This chapter offers a more epistemologically informed ethical analysis of how translational research teams might interact with various publics.

Chapter 10 pulls together various themes presented in this book, anchoring the narrative around the ethical difficulties presented by uncertainty. I then advance a series of research agendas extending from my analysis.

The issues explored in this book are largely situated at the juncture between the laboratory and the clinic. One consequence of this focus is that I will not explore with great depth a number of ethical and social questions traditionally associated with gene transfer. These include the deliberate inheritable genetic modification, the application of gene transfer to embryos and fetuses, and the use of gene transfer for cosmetic or enhancement interventions. I have elsewhere argued that the ethical questions raised by all three are no longer abstract or speculative: in vitro fertilization (IVF) clinics outside the USA *are* practicing a blunt form of inheritable gene transfer through ooplasmic transfer; clinical researchers *are* running trials investigating "life-style" conditions like erectile dysfunction.[22] Still, these phenomena lead us away from the primary focus of this book, which can be distilled to the question of how to best protect human subjects and the public where medical knowledge is at its most uncertain.

Obviously, a premise underwriting this book is that indeterminacy alters the ethical calculus surrounding human experiments. If so, a necessary first step for ensuring ethical research conduct would be for scientists and ethicists to recognize the profound uncertainty they confront when testing novel interventions in human beings for the first time. Most researchers probably need no reminder of this point. Nevertheless, I have witnessed numerous occasions of high-profile researchers offering blustery pronouncements that were demonstrably refuted by scientific developments a year or two later. In a letter to King Frederick William of Prussia in 1767, Voltaire wrote that "Le doute n'est pas une condition agréable, mais la certitude est absurde" (Doubt is unpleasant, but to be certain is absurd). The admonishment applies equally well to ethics. The issues discussed in the book abound with moral uncertainty, and the analysis offered here is intended not as the final word, but as an invitation for more sustained social and ethical analysis.

References

1. Okean MR. Anderson gets 14 years for child molestation. *Pasadena Star News* 2007 February 3.
2. FitzGerald S, Smith VA. Penn to pay $517,000 in gene therapy death. *Philadelphia Inquirer* 2005 February 10.
3. Beutler E. The Cline affair. *Mol Ther* 2001; **4**(5): 396–7.
4. Weiss R. NIH not told of deaths. *Washington Post* 1999 November 3.
5. Orkin SH, Motulsky AG. *Report and Recommendations of the Panel to Assess the NIH Investment in Research on Gene Therapy*. National Institutes of Health, 1995.
6. Friend T. It's in the genes: Scientists confront money issues. *USA Today* 2000 February 22.
7. Naldini L. Inserting optimism into gene therapy. *Nat Med* 2006; **12**(4): 386–8.
8. Xin H. Chinese gene therapy. Gendicine's efficacy: hard to translate. *Science* 2006; **314**(5803): 1233.
9. Guo J, Xin H. Chinese gene therapy. Splicing out the West? *Science* 2006; **314**(5803): 1232–5.
10. President's Commission for the Study of Ethical Problems in Medicine and Biomedical and Behavioral Research. *Splicing Life: A Report on the Social and Ethical Issues of Genetic Engineering with Human Beings*. US Government Printing Office, 1982; 126.
11. Walters L, Palmer JG. *The Ethics of Human Gene Therapy*. New York, Oxford University Press, 1997; 36.
12. The National Commission for the Protection of Human Subjects of Biomedical, and Behavioral Research. *The Belmont Report: Ethical Principles and Guidelines for the Protection of Human Subjects of Research*. US Department of Health, Education, and Welfare, 1979.
13. Jayson G, Harris J. How participants in cancer trials are chosen: ethics and conflicting interests. *Nat Rev Cancer* 2006; **6**(4): 330–6.

14. Richter R. $25 million gift to support bench-to-bedside research at Stanford. *Business Wire* 2006 September 14.
15. $13 million grant creates Wallace H. Coulter Center for Translational Research at University of Miami Miller School of Medicine. *University of Miami News* 2005 December 16.
16. The Translational Research Facility: Where collaboration leads to improved human health. *UMN News* 2004 Winter.
17. Kaiser J. Biomedicine. NIH funds a dozen 'homes' for translational research. *Science* 2006; **314**(5797): 237.
18. Zerhouni EA. Translational and clinical science – time for a new vision. *N Engl J Med* 2005; **353**(15): 1621–3.
19. Rowett L. U.K. Initiative to boost translational research. *J Natl Cancer Inst* 2002; **94**(10): 715–16.
20. Wadman M. Drive for drugs leads to baby clinical trials. *Nature* 2006; **440**(7083): 406–7.
21. Lehmann F, Lacombe D, Therasse P, Eggermont AM. Integration of translational research in the European Organization for Research and Treatment of Cancer Research (EORTC) clinical trial cooperative group mechanisms. *J Transl Med* 2003; **1**(1): 2.
22. Kimmelman J. The ethics of human gene transfer. *Nat Rev Genet* 2008; **9**(3): 239–44.

Chapter 2 What is gene transfer?

Introduction

Gene transfer researchers and others with historical inclinations like to describe an experiment performed by virologist Stanfield Rogers in the early 1970s as the first attempted gene transfer in human subjects.[1–4] The Oak Ridge National Laboratories scientist had observed that Shope papilloma virus infections, which normally cause warts, also depress blood serum levels of the amino acid arginine in rabbits. Studies also showed that workers handling the virus also had lower serum arginine.[5] Rogers then postulated that the virus contains a gene that codes for the enzyme arginase, which breaks down arginine.

Around the same time the German physician H. G. Terheggen in Cologne encountered a series of patients with various neurological impairments due to a deficiency in the enzyme arginase.[6,7] On learning of the report, Rogers contacted Terheggen and proposed administering the virus to patients with the enzyme deficiency. The study, notwithstanding its bold vision, proved unsuccessful in either improving disease symptoms or in producing biological insights. In fact, the virus used in the experiment "degenerat[ed]… in storage,"[8] and much later studies revealed that in fact the virus did not encode arginase after all.

There is, of course, a sense in which the conceit behind contemporary gene transfer can find precedent in this episode. But Rogers's experiment predated recombinant DNA technologies, which emerged in the mid 1970s and enabled manipulation of genetic sequences. It also predated the development of a biotechnology industry, or the knowledge economy, or the emergence of the "triple helix" configuration of universities, the private sector, and the government.[9] And in their wholesale abandonment of the strategy, Rogers's experiment stands in stark contrast to the perseverant and programmatic orientation of gene transfer research today. In this sense, Rogers's "first gene-transfer experiment" bears as much relation to contemporary gene transfer as semaphore does to modern telecommunications.*

So what is gene transfer? Scholar Sheila Jasanoff distinguishes between three different ways that biotechnologies have been framed in policy: as products, processes, and programs.[10,11] The first is the most familiar, and most accounts of the ethics of gene transfer begin with a description of the hardware. This book will not depart from that tradition. But a comprehensive account of gene transfer policy and ethics should also consider the processes by which these technologies are developed and applied, as well as how gene transfer fits within broader economic and social agendas. To that end, this chapter approaches the question "what is gene transfer?" from three different standpoints.

---

* At any rate, numerous earlier "gene-transfer" precursors can be identified. They include the use of oncolytic viruses in cancer treatment (traceable to the turn of the 20th century); earliest use of blood transfusion to treat hemophilia (1840s); and the development of organ and bone marrow transplantation throughout the 1950s and 1960s.

The product

I define human gene transfer as follows: *the use of genetic materials, genetic-based strategies, or genetically modified organisms to study or modify human biology.* This is more generic than conventional definitions, which tend to use the term "gene therapy" to mean "the use of normal genes or genetic material to replace or cancel out the 'bad' or defective genes in a person's body that are responsible for a disease or medical problem."[12†]

There are many reasons why the conventional terminology and definition are unsatisfactory. One is that that, for years, many in the research community have called these techniques "gene therapy" despite a lack of evidence of efficacy. Because this optimistic characterization of the technology can cloud risk assessment, public communications, and informed consent, this book will use more the neutral terminology of "gene transfer." In addition, the conventional definition lacks comprehensiveness. For example, some of the most informative "gene therapy" studies in the first five years of clinical testing involved "gene marking," in which cells were tagged with genetic sequences in order to determine whether certain cells proliferated in the human body, or where they traveled. Lastly, one of the most controversial extensions of "gene therapy" is genetic enhancement (that is, the use of gene transfer technologies towards cosmetic ends such as increased muscle mass for athletes, or higher intelligence in children). Such applications fit very uncomfortably under the category of "therapy." According to Sheldon Krimsky, the term "gene therapy" originated as a kind of lexical foil to 1980s controversies surrounding human genetic engineering.[18] Regardless of whether this account is historically accurate, I think it is safe to say that "gene therapy" obscures many of the central ethical issues surrounding the techniques and practices it embodies (the same goes for "cell therapy," which I prefer to call "cell transplantation," and for "therapeutic cloning," which I prefer to call "somatic cell nuclear transfer.")

Another reason to question conventional definitions is that there is much they leave out. This will be apparent upon the most casual browsing of gene transfer journals like *Molecular Therapy*. For example, researchers are currently developing genetically modified viruses aimed at stimulating powerful immune responses to infectious disease agents like HIV. Others are attempting to genetically modify intestinal bacteria to secrete growth factors in order to treat conditions like Crohn's disease.[19,20] Another promising technique is modifying viruses so that they selectively infect and kill tumor tissues. Such strategies do not involve modifying genes in a person's body, unless one counts unwelcome viruses, or gut flora, as part

† The US Department of Energy defines gene therapy as "a technique for correcting defective genes responsible for disease development. Researchers may use one of several approaches for correcting faulty genes."[13] The official definition of the American Society of Gene Therapy, the US professional society, is "an approach to treating disease by either modifying the expressions of an individual's genes or correction of abnormal genes."[14] Elsewhere, this organization uses a slightly modified definition: "the process of inserting nucleic acids (e.g. usually DNA/genes) into cells or tissues to correct or prevent a pathological process."[15] The European Society of Gene and Cell Therapy uses the following definition: "a technology by which genes or small DNA or RNA molecules are delivered to human cells, tissues or organs to correct a genetic defect, or to provide new therapeutic functions for the ultimate purpose of preventing or treating diseases."[16] The OECD definition is: "Gene delivery, the insertion of genes (e.g. via retroviral vectors) into selected cells in the body in order to: Cause those cells to produce specific therapeutic agents; Cause those cells to become (more) susceptible to a conventional therapeutic agent that previously was ineffective against that particular condition/disease; Cause those cells to become less susceptible to a conventional therapeutic agent; Counter the effects of abnormal (damaged) tumour suppressor genes via insertion of normal tumour suppressor genes; Cause expression of ribozymes that cleave oncogenes (cancer-causing genes); Introduce other therapeutics into cells.,"[17]

of a person's body. Other strategies do more than simply "replace" or "cancel out" genes in a person's body. For example, one strategy researchers have pursued involves genetically modifying stem cells that give rise to blood tissues in order to enable patients to withstand higher doses of chemotherapy.

Another factor that complicates the definition of gene transfer is the question: what is a gene? Early on, gene transfer experiments used genetic sequences that clearly fit the canonical definition of "gene": a sequence of nucleotides that encodes a protein product. Quickly thereafter, however, scientists began pursuing oligonucleotide strategies, in which chemically modified DNA sequences are administered to a person to thwart the production of certain gene products. More recently, many researchers have been pursuing RNA interference approaches, in which small genetic sequences are used to produce short RNA molecules that bind to specific RNA sequences in a person's body, thereby "knocking down" expression of genes. These interfering RNAs do, indeed, count as "genes," but not in the traditional sense.

The target

The techniques and strategies of gene transfer are immensely varied. Nevertheless, they sort into several politically and ethically relevant classes. One key distinction is the entity to which genetic materials are transferred. These divide into three categories: germ cells, somatic cells, and foreign organisms.

Germ cells produce sperm or ova. Genetic modifications of these cells will generally be passed on to the recipient's progeny. Inheritable genetic modification has received considerable attention from ethicists and policy makers. Many jurisdictions, like the European Community,[21] Canada,[22] and India,[23] ban the practice. Officially, the technique is not currently being pursued in human beings (though germ cell modification is being investigated preclinically). "Unofficially," however, some infertility clinics outside of Europe and North America do offer ooplasmic transfer, which involves transplanting mitochondria from viable ova into those of infertile women.[24] Because mitochondira contain their own genome, the technique is arguably an oblique form of germline modification. Readers interested in the ethics of germ cell modification are directed to other sources.[25,26]

Somatic cells include any tissues that do not give rise to sperm or ova; they include skin, liver, blood, the nervous system, muscle, tumor cells, etc. The therapeutic benefits or risks of somatic gene transfer stop with the individual recipient of the gene transfer. Somatic cells can, themselves, be divided into two broad categories: those that are highly differentiated and those that are not (stem cells). Differentiated tissues do not develop into different tissue types and, as a rule, do not propagate. If genetically modified, differentiated tissues are often lost to aging or injury. In contrast, stem cells in adults can develop into different tissues and replenish themselves. Gene transfer aimed at adult stem cells (and attendant risks and benefits) is thus more likely to be permanent.

Modified foreign organisms include bacteria, viruses, or fungi. These were briefly mentioned above. Though the target here is the individual patient, one concern these approaches raise is the possibility that modified organisms will spread to members of the public.

Vehicles and vectors

With millions of years of evolutionary pressure, mammalian organisms like human beings have developed numerous ways to safeguard their genomes and repel foreign DNA. One of the central challenges for gene transfer, then, is delivering genetic materials into a person's cells and getting those materials to express themselves.

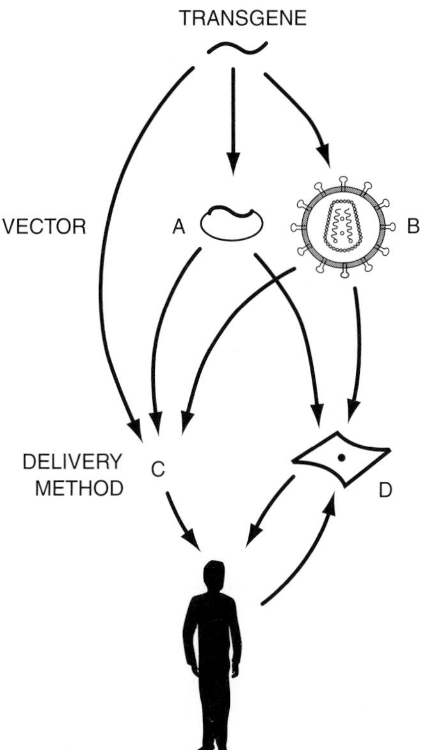

Figure 2.1 Gene transfer strategies. Transgene can be either delivered directly to the human being, or inserted into a vector. Vectors take two basic forms: nonviral (A) and viral (B). Transgenes or vectors can then be delivered through two basic routes: in vivo (C), in which vectors or transgenes are delivered directly to the patient, or ex vivo (D), in which tissues are extracted from a patient, modified in a laboratory, and then returned to the patient.

Delivery strategies divide into two categories (shown schematically in Figure 2.1). One approach is to simply inject genetic material into a human being. Sometimes, researchers package genetic materials in a substance, for example a cationic lipid, that improves the ability of genetic materials to penetrate cell membranes. These approaches, however, are uniformly inefficient. Those cells that do take up genetic material retain it for a short time. Collectively, these approaches are referred to as "nonviral" gene transfer, and they account for a quarter of all gene transfer trials.

The other approach uses various viruses ("viral vectors") to infiltrate human cells and deposit genetic materials (the proportions of trials using different strategies are shown in Figure 2.2). In the vast majority of viral vector studies, viruses are disabled so that they cannot propagate. According to a global database of gene transfer clinical trials, the most common viral vector is the adenovirus. Adenoviruses normally cause acute respiratory infections, and account for about 10 percent of common colds in children. However, most adenoviral vectors used in gene-transfer studies have had many (if not all) of their genes removed so that they are incapable of establishing an infection. Their protein coats nevertheless can provoke powerful immune responses. Adenoviruses are generally good at entering cells, and producing very large quantities of an intended gene product (termed "transgene product"). On the other hand, apart from exceptional circumstances, genetic materials transferred by adenoviral vectors are usually lost with time and, as will be described in Chapter 3, they can cause fatal immune reactions.

The second most common vector is retrovirus based. A key advantage of retroviruses is that they can integrate their genetic materials into the host genome. As such, retroviral gene transfer – unlike adenoviral gene transfer – is permanent. There are two problems with

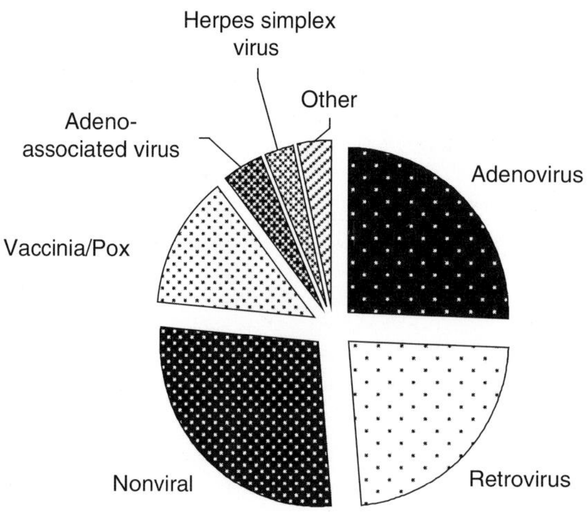

Figure 2.2 Vectors used in gene transfer. Derived from Wiley, *Journal of Gene Medicine*, "Gene Therapy Clinical Trials Worldwide" (www.wiley.co.uk/genmed/clinical/; accessed August, 2008).

retroviral vectors. The first is that they are incapable of entering the nucleus of non-dividing cells (like neurons or mature tissue). The second is that, because there is no way to control where they integrate, there is a risk that vectors will activate oncogenes or disrupt tumor suppressor genes, leading to cancer.

One subcategory of retroviruses that is receiving increased attention is the lentivirus, the most infamous example of which is HIV. Lentiviruses have many of the advantages of other vectors: they accept large DNA inserts, and integrate. A major advantage over other retroviruses is that they can infect non-dividing tissues. There is also some suggestion that they may be less prone to causing cancers. Nevertheless, the possibility of oncogenesis cannot be ruled out; nor can biosafety concerns surrounding manipulation of vectors that are closely related to a human scourge.

Another major viral gene transfer category is adeno-associated virus (AAV) vectors. Viruses from which these are derived are generally considered to be harmless (there is conflicting evidence concerning whether these vectors can cause cancer; so far, however, they have not been associated with clinically significant side effects). AAV vectors are capable of integration, though they are more typically maintained outside host chromosomes (that is, episomally) for a long period. The main problem with AAV vectors is that they can only transfer short genetic sequences. They are also somewhat difficult to manufacture, though production techniques have improved dramatically over the last decade.

Vaccinia and pox virus studies account for the third largest category of vector strategies used in gene transfer. These viruses are typically used in vaccination studies to provoke immune responses against proteins on the surface of tumor cells. Finally, there are a number of other viral vectors that are currently being explored. One is the herpes simplex virus (HSV). Others vectors in human studies are derived from measles, Newcastle disease, sendai, rheovirus, simian foamy virus, semliki, and vesicular stomatitis viruses.

Viruses come in different serotypes, that is, genetic varieties. For example, nine serotypes of AAV have been identified. Different serotypes often have properties that make them attractive for some applications, and unattractive for others. Thus, AAV8 vectors delivered to the liver result in more rapid expression of transgene than AAV2; AAV1 vectors are more efficient at transducing central nervous system tissues than AAV2, but AAV2 is generally the best vector for transfecting kidney tissue.[27]

Delivery approaches

Another major challenge in gene transfer is delivery. On the one hand, getting vectors or vehicles to the appropriate tissue can be exceedingly difficult, particularly if the tissues are inaccessible (e.g. the brain). On the other hand, vectors can spread throughout the body, and transfer genes to inappropriate tissues.

Delivery approaches divide into two categories. In vivo approaches administer vectors directly to the volunteer, for example, through injection into the bloodstream, muscle, or a tumor, or as a mist delivered to the lung. Scientists will sometimes modify vectors so that they target specific tissues by adding proteins that recognize receptors on particular cells. The problem with in vivo approaches is twofold. First, vectors can reach unintended targets, causing transgenes to be expressed in inappropriate tissues. Second, systemic administration can cause strong immune reactions.

In some circumstances, in vivo gene transfer protocols require relatively novel surgical procedures. For example, one trial studied the effects of administering adenoviral vectors to the eye in patients with the rare eye cancer retinoblastoma. The injection procedure raised concerns that needle tracks left behind would provide a passage for retinoblastoma tumors to spread outside the eye[28] (fortunately, this did not occur[29]).

Ex vivo approaches involve removing tissue from a person, transferring genetic materials in the laboratory, and then transplanting cells or tissues into a human being. Many such protocols use bone marrow-derived tissues. Ex vivo approaches tend to involve greater control over where the vectors end up in the body ("biodistribution"). On the other hand, such procedures tend to be more cumbersome and costly than direct approaches.

The genetic material

There is virtually no end to the types of genetic material that are being tested in human studies. One broad category is gene replacement, which involves giving functional genes to overcome deleterious genetic mutations in a patient. These include coagulant factor IX for hemophilia patients and cystic fibrosis transmembrane receptor in cystic fibrosis patients, or dystrophin for patients with muscular dystrophy.

Gene addition comprises another broad category; this involves the transfer of genes that are not normally (or adequately) expressed in a particular tissue. The types of genes that might be added include growth factors and cytokines (to stimulate tissue growth or immune reactions), antigens (to trigger an immune response), or small genes like oligonucleotides or siRNA (to inhibit expression of a disease-causing gene). About 8 percent of human studies involve addition of suicide genes – that is, genes whose products convert relatively benign pro-drugs into toxins that destroy cancer cells; another 4 percent of protocols involve marking – that is, tagging tissues with inert genes that can be used to track the fate of cells.[30]

Just another drug?

An often heard suggestion is that gene transfer is just another drug technology. Chapter 4 will explain why, at least with respect to safety, the drug analogy confounds as much as it reveals. For now, I note a few key differences with respect to the product definition. First, their composition is complex. Most drugs are relatively simple in terms of their molecular characteristics. Gene-transfer agents are mixtures, consisting of multiple genes (e.g. transgene, promoters, viral elements, selectable markers) as well as other design features required for function (e.g. various capsid proteins, other viral proteins). There are no simple ways to determine with precision the contents of the vial. For example, vector preparations contain a cocktail of vector particles (some of which are empty or otherwise deficient), and their genetic elements might acquire mutations in the course of production.

Another key difference is the multiplying diversity of platforms. Just as the term "cancer" conceals the variety of disease processes, so too does gene transfer contain multitudes. The effects and properties of ex vivo approaches will vary depending on the characteristics of cells. Other strategies being explored include nanotechnology approaches, artificial chromosomes, new vectors and serotypes, and vectors that carry multiple genetic elements.

A third key difference is that many gene transfer products are transformed in the course of application. For instance, ex vivo approaches require harvesting and alteration of tissue, followed by transplantation. Each step alters the composition and characteristics of the product, introducing new ways in which interventions can fail.

The process

Contemporary gene transfer is characterized by a distinctive – if not strictly unique – configuration of various actors. Key to appreciating this configuration is the recognition that, as one gene transfer researcher put it in 2008, "no one's making any money yet." Despite numerous promising clinical results, gene-transfer products have yet to be commercialized in North America, Japan, or Europe. As indicated in Figure 2.3, the volume of published gene-transfer studies that identify themselves as preclinical follow parallel trends to published clinical studies until 2000, after which the preclinical studies dramatically eclipse clinical studies in terms of volume. Another indication that laboratory research in this field currently dominates clinical research is the professional degrees of its practitioners: Ph.D. membership in the American Society of Gene Therapy exceeds that for M.D.s by a ratio of 3:1 (M. Dean, personal communication, 2009).

The vast majority of human studies involve small-scale trials aimed primarily at testing toxicity, dosages, and biological effects. According to a widely used database of gene transfer studies, 61 percent of human studies are phase 1 trials; another 20 percent of studies are combined phase 1/phase 2 trials (see Figure 2.4; throughout this book I refer to pilot, phase 1, and combined phase 1/phase 2 studies as "early-phase" trials).

What accounts for the peculiar arc of clinical trial volume, which declines sharply in 2000 and plateaus two years later? Several explanations can be offered. The "naturalistic" explanation is that this is typical for novel interventions, which tend to be introduced into human testing in an environment of exuberant expectation, after which ambitions are scaled back as limitations are discovered. This explanation will be examined critically in Chapter 9. For now, however, I note that while plateaus in trial volumes of novel therapeutics are indeed recurrent, declining volumes have tended to occur only where there is widespread recognition of insuperable obstacles to progress (for example, organ transplant surgeons enacted moratoria on recognition of immunological barriers). Another explanation is the death of a medically stable, 18-year-old man in a gene-transfer study conducted at the University of Pennsylvania (discussed in Chapter 3). Not long after, leukemias in subjects in a French study involving X-SCID (see Chapter 4) further stalled interest in the technique. A third explanation – not wholly independent of the previous two – is a shift in the economic fortunes of gene transfer. Year 2000 coincides with significant retrenchment on the part of the biotechnology sector. Indeed, a chart of a biotechnology stock index is congruent with that of gene transfer volume. It's possible, then, that withdrawal of biotechnology support from gene transfer precipitated a sharp decline of trial volume, even in the public sector.

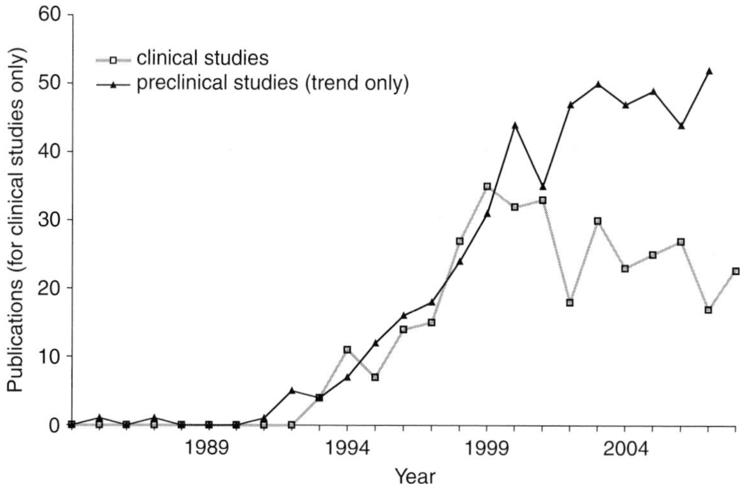

Figure 2.3 Trends in volume of published gene transfer clinical trials and preclinical studies. The graph mainly illustrates trends, not absolute figures. In particular, numbers in the y-axis apply only to clinical studies. Method: *PubMed* was searched on October 6, 2008, using the following strategy: clinical studies: ("gene transfer" OR "gene therapy") AND (pilot OR marking OR "phase I" OR "phase 1" OR "phase 2" OR "phase 3") LIMITS: clinical trials; list was then hand searched to exclude reviews and protocols. Preclinical studies: ("gene transfer" OR "gene therapy") AND (preclinical OR "pre-clinical") NOT review NOT advances NOT production NOT manufacture LIMITS: animals. The numbers provided for clinical studies represent a rough estimate of actual studies, though these figures are subject to the limitation that any given gene transfer trial might produce multiple publications (e.g. preliminary reports, phase 1 study, long-term follow-up study, and studies using biobanked tissues from trial), and some trials go unpublished. The numbers provided for preclinical studies are crude because many preclinical studies might not describe themselves as such in the abstract.

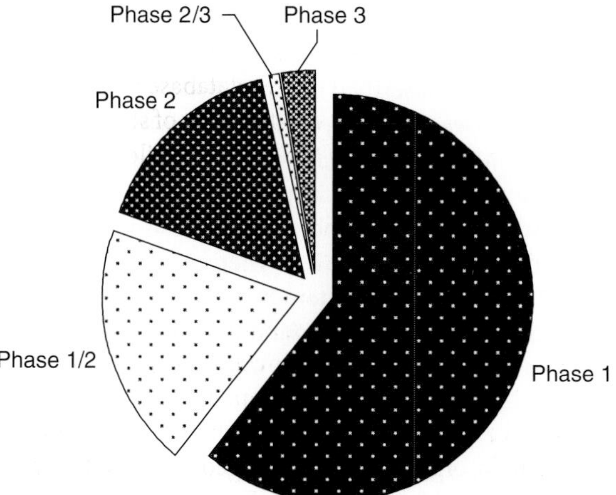

Figure 2.4 Phases of gene transfer clinical trials. Derived from Wiley, *Journal of Gene Medicine*, "Gene Therapy Clinical Trials Worldwide" (www.wiley.co.uk/genmed/clinical/; accessed July 19, 2008).

The actors

Five major interest groups shape the field of human gene transfer: clinicians and laboratory scientists, sponsors, institutional hosts, disease advocates and patients, and regulators and ethicists.

The scientists and clinicians in gene transfer, like other translational researchers, occupy a boundary zone between the basic laboratory and the medical clinic. Trials tend to be large undertakings, drawing on a diverse range of expertise. For example, the hemophilia study referenced in the previous chapter named 31 authors with 15 different institutional affiliations, on three continents.[31] Lead clinical investigators often commute between clinical work and laboratory research, and are involved in originating and developing agents tested in clinical trials. The prominent role that inventors play in bringing their products to clinical trials is indicated in the frequency with which principal investigators are named as inventors on products that relate closely to study interventions.

Conducting a career in the liminal space between the clinic and the laboratory is fraught with challenge and tension. For one thing, clinical research is notoriously expensive. At the same time, because of ethical restrictions on human experimentation, study designs in clinical research tend to be less elegant, and winning funding from public agencies is no small task. Further aggravating funding difficulties is the placement of trial results – which take years to collect and rarely confirm clinical promise on first attempt – in high-impact journals. That said, numerous negative or ambiguous gene-transfer trials have been published in prominent journals.[32]

Translational researchers also find themselves caught between the reward cultures of medicine and science. The clinician secures authority among their watchful physician colleagues, regulators, patients, and ethics review bodies by projecting certainty, confidence, and closure. Thus, they seek to persuade various parties that safety is assured to a reasonably high degree, and that they have exhausted avenues for producing further knowledge through preclinical and basic testing. On the other hand, scientists secure authority by exposing uncertainty and opening up new lines of inquiry.[‡] As a consequence, basic scientists commonly view the confidence of translational researchers with some suspicion.

One characteristic of gene transfer is the degree to which it has constituted itself as a distinct field of biomedical inquiry and practice. As described by Christine Crofts and Sheldon Krimsky, numerous journals with names like *Gene Therapy*, *Cancer Gene Therapy*, *Molecular Therapy*, *Human Gene Therapy*, and *Journal of Gene Medicine* are devoted to reporting developments in gene transfer; many have reasonably high impact factors, an indicator that papers they publish are cited by other authors. US researchers have established a large professional society for gene transfer, the American Society of Gene Therapy, as have Europeans (European Society for Gene Therapy). This platform-centered organizing contrasts with other scientific and medical fields, which tend to revolve around diseases or medical specialties. Thus, journals specifically devoted to platforms like monoclonal antibodies or

‡ A good illustration of this tension can be witnessed in the RAC deliberations surrounding the hemophilia trial profiled in Chapter 1. The researchers had proposed the trial before having completed a similar study in which vector was delivered to muscle, and a non-human primate study of vector safety. On the one hand, the researchers had to convince the committee that the muscle trial adequately demonstrated the safety of administering AAV vectors to volunteers with hemophilia. To support the claim of a favorable risk–benefit balance for hemophilia patients who might enter the study, they also had to use preliminary findings from their muscle study to support the claim that AAV vectors were adequately promising. But they also had to use these same promising findings to project uncertainty about the prospect of AAV gene transfer for hemophilia, uncertainty that could only be resolved by testing liver administration of vector.[33]

recombinant proteins are scarce or generally of lower impact, as are corresponding major professional societies.[§]

Though there are no databases for gene-transfer study sponsorship, the main source of funding in North America – be it clinical or preclinical, direct or indirect – is the public sector. A search of a comprehensive database of clinical trials, clincaltrials.gov, shows that 54 percent of phase 1 gene-transfer trials receive direct funding from the NIH; 67 percent receive funding from universities or foundations (many studies are co-funded). Between 2004 and 2007, the National Institutes of Health provided \$325–391 million annually for gene-transfer research, and another \$31–37 million for gene-transfer trials.[34] One of the largest institutional sponsors of early-phase gene-transfer trials within the NIH is the National Institute of Cancer (NCI). From 2000 to 2008 inclusive, the NCI funded 502 projects involving phase 1 gene-transfer trials (to put this in perspective, the figure is approximately equivalent for cancer vaccine studies, half that for molecularly targeted therapies, and one fifth that for chemotherapy).[35] Still, NIH has lavished support for some institutions and investigators pursuing gene transfer. In 2001, *Science* magazine published a listing of top-funded basic research PIs; three were gene transfer researchers.[36]

Another major source of funding – though perhaps not as important as public perception would have it – is the private sector. Over the most recent two years, only two phase 1 trials were funded by large pharmaceutical companies. Instead, the biotechnology industry is the main commercial actor for translational trials. Though several companies have survived a decade of gene-transfer research, the number of gene-transfer biotechnology ventures has tended to fluctuate from year to year.[37] From 2006 to mid 2008, fifteen phase 1 gene-transfer trials were sponsored by biotechnology companies that were less than ten years old; six of these were less than five.[¶]

Despite their prominence in several high-profile controversies,[38,39] commercial interests have never really dominated the field. Only 33 percent of all US gene-transfer trials registered with the NIH identify private sector sponsors[40] (the clinical trials database of more recent trials shows that only 21 percent of phase 1 studies receive industry support).[41] One presumes, then, that the remaining two-thirds is investigator initiated, and covered through NIH grants, foundation support, and/or operating budgets. The inference that private sponsorship of phase 1 studies is not primary is further supported by an examination of acknowledgments sections of phase 1 clinical trials published over the last two years. Approximately one-fifth report funding from the private sector. In contrast, 64 percent acknowledge funding from government sources, and 36 percent are supported by foundations.[**]

Academic medical centers are a major source of financial and operational support in early-phase gene-transfer trials. Almost the same proportion of gene-transfer trials that acknowledge biotechnology company support also name academic research centers as sponsors. One of the

[§] There are, of course, exceptions. For example, cancer immunotherapy has at least two journals and a professional society, though neither would seem to have the profile of their equivalents in gene transfer. One field that has constituted itself in a way similar to gene transfer is cell therapy, which has established several prominent journals and at least one large professional society, ICCST.

[¶] These estimates were determined on the basis of founding dates for sponsors of most recent 100 registered phase 1 trials with RAC. For two companies, founding dates were not available.

[**] Based on a *PubMed* search of phase 1 studies, performed on June 20, 2008. The total sample included 22 published studies. Preclinical or phase 2 studies were excluded, as were studies outside of Europe, North America, and Japan. In three instances, studies did not contain any acknowledgment of funding sources. Percentages do not add to 100 because several studies report multiple funding sources.

first centers was established at the University of Pennsylvania in 1992;[42] since then, numerous other medical centers around the globe now have their own programs, which variously provide vector production facilities, clinical research services, toxicology testing, regulatory affairs support, and training.

One last key source of support is private foundations. Where there are highly morbid and mortal diseases affecting Americans and Europeans, there are also scientifically and politically sophisticated organizations funding and advocating cutting-edge research. Patient activists and disease philanthropies have played important roles in early-phase gene transfer as sponsors, advocates, and information clearing houses. Concerning the first, support for gene-transfer research has tended to cluster among charities that advocate on behalf of persons with rare genetic disorders. Thus, for example, American Society of Gene Therapy annual meetings have consistently drawn financial support from groups like the March of Dimes and the Fanconi Anemia Research Fund.

Disease advocates have also helped shape policy and regulation at key junctures. In numerous instances, disease foundations have lent support for researchers or mobilized patient testimony for studies that received skeptical hearings on regulatory review. Disease foundations have tended to advocate a more permissive policy for early-phase studies. A perhaps extreme illustration of the views that some disease advocates have towards regulatory and ethical review can be found in Phil and Tricia Milto's statement on their son's struggle with Batten disease: "Many times, potential therapies for rare diseases do not make it through the approval process because of the Recombinant Advisory Committee (RAC). The RAC is a politically charged group that oftentimes denies approval of innovative therapies by requiring more testing when it is not required under the normal FDA guidelines for orphan diseases, costing precious time and money. These potential therapies are the only chance for life for children with rare diseases. Children die because of regulatory issues. This is a well-documented problem that all rare diseases face."[43]

Last, disease foundations have often provided important organizational and administrative support for gene-transfer studies. In terms of enabling trials, they provide researchers with a mechanism for advertising studies and recruiting volunteers. They also help shape the design of studies by coordinating efforts and accelerating the flow of information across different research sites. Thus, for example, the UK Cystic Fibrosis Trust has established a consortium of cystic fibrosis gene-transfer researchers consisting of 80 members, while in the US, the Cystic Fibrosis Foundation provides much of the support for a network of clinical research activities.[44]

Patients in gene transfer represent a heterogeneous class in all respects except one: few are healthy. The vast majority of early-phase studies involve patients who have exhausted standard treatment options (in the language of medicine, they are "treatment-refractory"). In general, participants in phase 1 cancer studies are more likely to be educated, white, and affluent.[45,46] Patients offered participation in phase 1 cancer studies – if they meet eligibility criteria – generally enroll. According to one recent study, only 19 percent of eligible patients declined enrollment in a phase 1 cancer study.[47] Though these figures are derived from phase 1 cancer trials, which consist mainly of small-molecule drug studies, there is no reason to believe that these figures would be any different for gene transfer.

In specialty areas like cystic fibrosis, volunteers often have extensive experience with research participation and can be highly informed about gene transfer and the trial process.[48,49] Nevertheless, gene-transfer trials also frequently target infants and children – a group that is vulnerable – and, as will be explored in Chapter 8, economically disadvantaged groups are also occasionally recruited to specific studies.

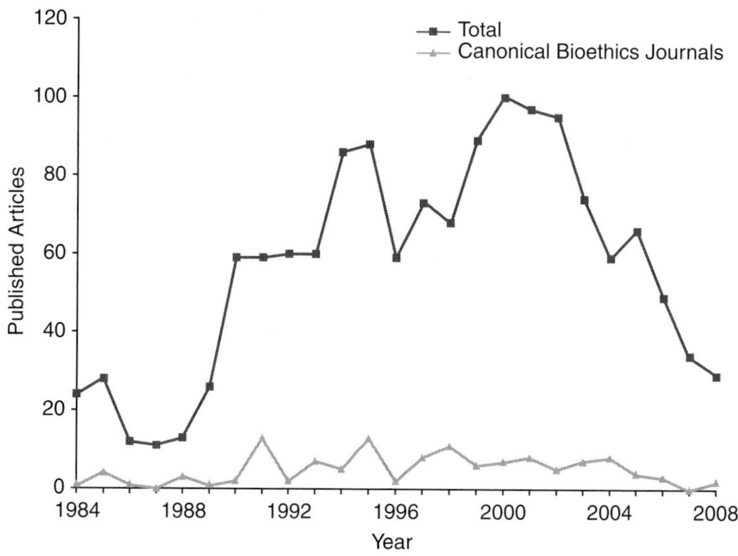

Figure 2.5 Volume of scholarly articles on gene transfer oversight and ethics. Method: For total, *PubMed* was searched using search terms gene therapy OR gene transfer; article subsets were limited to Bioethics; to exclude news reports, article types included were: addresses, classical article, comment, consensus development conference, NIH, corrected and republished article, English abstract, evaluation studies, festschrift, guideline, historical article, interview, journal article, lectures, legal cases, multicenter study, overall, published erratum, research support, NIH extramural research support, NIH intramural research support, Non-US Gov't Research Support, U S Gov't, non-PHS research support, US gov't PHS, scientific integrity review, technical report. Non-English articles were also excluded. For canonical bioethics journals, the total dataset was searched for articles published in the following journals: *Hastings Center Report; Cambridge Quarterly of Healthcare Ethics; Journal of Medicine and Philosophy; IRB; Journal of Law, Medicine & Ethics; Bioethics; Journal of Medical Ethics; American Journal of Bioethics.*

Regulators and ethicists represent the final major actors in gene transfer. The former will be addressed further below. The latter consist of a range of commentators – many of them philosophers and social scientists – who hold academic appointments at medical centers. The volume of scholarly literature on gene transfer oversight and ethics is astounding (see Figure 2.5), and publications easily outnumber actual trials. The volume of publications has fallen off much more steeply than has that of trials, though the drop is coincident with trial volume. As indicated in Figure 2.6, commentary and scholarship have tended to focus on risk and reproduction, though with time, these two issues have receded as a focus.

The assimilation of ethics and regulation to gene transfer is apparent in the layers of regulatory and ethical oversight for trials, the existence of informed consent guidelines specific to gene transfer,[50] the establishment of ethics-related policies by major gene-transfer societies, and the abundance of ethics- and policy-related commentary appearing in journals that primarily target gene-transfer researchers.[††] For many in the field, this attention is unwelcome and viewed as inhibitory. In many instances, external ethical and regulatory bodies have quashed studies or barred activities like in utero gene transfer. Nevertheless, I would suggest that there are a number of ways in which ethics has helped shape gene

[††] From 1998 to mid 2008, almost 100 bioethics articles, comments, or editorials have been published in the premier gene-transfer speciality journals *Human Gene Therapy* and *Molecular Therapy*. [*PubMed* advanced search on June 25, 2008; Topic limited to Bioethics]

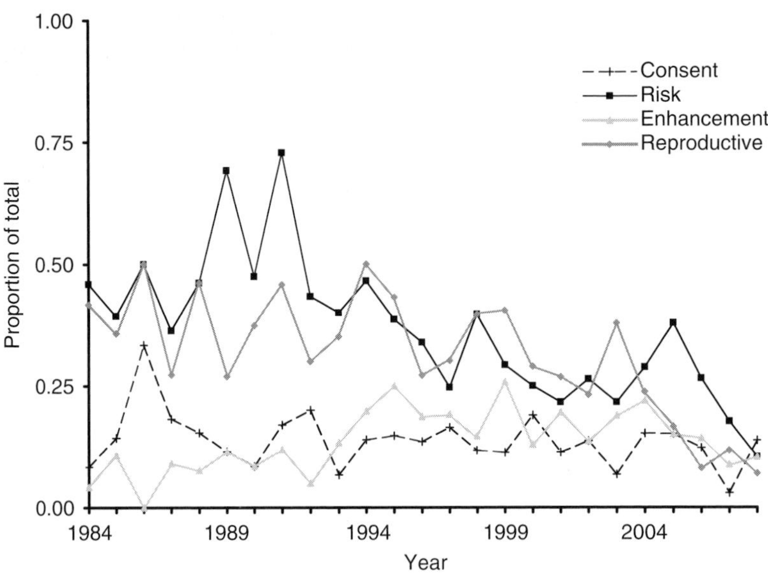

Figure 2.6 Focus of ethical analysis in gene transfer. Method: total database from Figure 2.5 was searched using the following terms: Consent: "consent." Risk: "risk OR safety." Enhancement: "enhance* OR cosmetic." Reproductive: "germ OR germ* OR inherit* OR utero OR fetal."

transfer. One way concerns how ethics "tags" certain issues as flashpoints for deliberation. In his book, *Playing God?*, sociologist John Evans describes the emergence of an ethical discourse that all but excluded discussion of the ends of human genetic engineering.[51] A consequence, according to Evans, is that clinical researchers seldom need worry about defending the objectives and ultimate applications of their research programs.[‡‡] Another issue concerns benefits: the last two decades of gene transfer have, if anything, demonstrated the profoundly distributed nature of biomedical innovation, as well as the large commitment of public resources necessary to actualize clinical applications. How will access be determined? In the USA, the pricing of biological products and regulatory hurdles to the licensure of generic (or "follow-on") biologicals provide grounds for worry that the fruits of this large public endeavor will not be shared evenly. Such concerns, however, are bracketed from ethical discussion.

A related way that ethics has helped smooth the path for gene transfer is by shaping uncertainty. When gene transfer was first proposed in the early 1980s, its public and policy reception was caught up in broad-ranging, poorly articulated concerns including whether genetic manipulation "usurped" a natural or spiritual order, commercial domination of research and development, eugenic implications, disability rights and interests, and public accountability of basic and clinical research. At least in Anglophone countries, however, bioethics has helped to center concerns largely around those related to product safety and informed consent in research. In a similar vein, when trials were first proposed, there was considerable uncertainty and debate about the appropriate criteria for launching studies. Through RAC review,

[‡‡] For example, the RAC currently restricts germ cell modification not for reasons of ends, but rather because of safety concerns. The presumption is that once safety concerns are eliminated, any genetic manipulation of human beings is permissible.

the pathway for approval of first-in-human trials has become clearer, and trialists wishing to pursue clinical testing confront less uncertainty. Some researchers, perhaps cynically, admit to having used RAC review as a way of establishing a clear, unambiguous goal towards which to direct laboratory activities.[52]

Third, ethics serves an instrumental social purpose, ensuring that research and its conduct are in equilibrium with consensus values. Put another way, ethics oversight helps keep research out of the morning papers. This is not to say that ethics does not occasionally create liabilities for researchers or uncover contentious conduct. Nevertheless, the suggestion is plausible that ethics oversight helps foster public impressions that prohibitions are unnecessary.

The categories of actors described in the previous paragraphs are often fluid. For instance, a former vice-president of research and clinical development at Avigen – a company that pursued gene-transfer trials against hemophilia – also served as President of the National Hemophilia Foundation. Parents of a child with Canavan disease at once established a charity to raise funds for phase 1 studies, and enrolled their son. Researchers have founded and served in the leadership of various biotechnology companies developing gene-transfer products. Academic medical centers have negotiated partnerships with biotechnology companies, and even operate their own "nonprofit" biotechnology start-ups.[53] Groups like the Cystic Fibrosis Foundation and Multiple Myeloma Research Foundation finance product development and, in certain respects, run like a biotechnology company.[54–56] Gene transfer researchers serve on oversight bodies like RAC, and ethicists (like myself) receive spillover funding from grants primarily directed towards gene-transfer development.

Oversight

In most countries, drug studies in human beings are regulated at two different levels. First, health agencies typically require submission of a protocol for clinical studies pursued with the goal of commercial licensure. In the USA, sponsors submit an "Investigational New Drug" application to the Food and Drug Administration. If the proposal meets with approval, the FDA will then issue a "no objection" letter that authorizes the investigation to proceed. Gene-transfer agents are generally regulated as biologics, a class that includes vaccines, tissues, blood products, agents purified from animals, and some recombinant proteins.[57] Note that, using my broad definition of gene transfer, some gene-transfer strategies, including oligonucleotidetherapies, are regulated instead as drugs.[58] Because gene transfer is characterized by rapid technological change, the FDA has tended to rely on flexible guidance and "points-to-consider" documents rather than full-blown regulations.[59] A similarly plastic approach to regulation is apparent with the International Council of Harmonization, which in 2001 established a "discussion group" to monitor the science of gene transfer and exchange information with the aim of harmonizing international regulations.[60]

Second, protocols undergo review by a research ethics committee (these go by different acronyms depending on the jurisdiction; the one that will be used in this book is IRB). In addition to considering risk and benefit, IRBs concern themselves with informed consent, fair subject selection, conflict of interest, publication agreements, etc. In many countries IRB review is a local affair, with universities forming committees that review experiments performed under the same roof. Both IRB and drug agency review are strictly confidential. As such, there are few opportunities for members of the public, other ethicists, and patient groups to learn about the substance of particular protocols, where reviewers had concerns, and whether or how researchers addressed these concerns.

For largely historical reasons, gene-transfer trials in many countries undergo two additional layers of review.[§§] Since the first official gene-transfer protocol in 1989, all US studies pursued at institutions that receive federal funding for research involving recombinant DNA must submit protocols for review to two other bodies.[¶¶] The first is an Institutional Biosafety Committee. IBCs are charged with reviewing measures to control the spread of genetically modified agents. Like IRBs, IBCs are local, though NIH policy stipulates that meeting minutes be made available to the public.[64] IBC activity has, until recently, received very little public scrutiny. The events of September 11th changed that. Surging interest in biodefense research, including a proliferation of laboratories equipped to handle dangerous pathogens like *Ebola* virus, brought increased attention to these bodies. According to activists Edward Hammond and Margaret Race, the IBC system is marked by serious deficiencies and non-compliance.[65] Does this pose a threat to public health, and are these threats deriving from gene-transfer research per se, or unrelated virology studies? I am not aware of any instances where laboratory personnel or members of the public have become infected during the course of gene-transfer trials, but there certainly are instances where workers have become infected with vectors very similar to those used in gene-transfer studies.[66]

The second unique layer of review for gene transfer is centralized review. In the USA investigators at institutions receiving NIH funding for recombinant DNA research must submit human gene-transfer studies for review by the Recombinant DNA Advisory Committee. Since 1993, the RAC has restricted full, public review to protocols raising novel concerns about ethics or safety.[62,64] Until 1996, the RAC enjoyed a kind of authority to approve or reject protocols by passing its recommendations on to the NIH director. Since reforms instituted by NIH director Harold Varmus, the committee was reorganized as an advisory body. Other countries, like the UK,[67] Australia,[68] and the Netherlands,[69,70] have analogous centralized review structures; Canada does not.***

Through the Freedom of Information Act, nearly all materials submitted to the RAC – and commentary by reviewers – are publicly accessible; since 2000, meetings have been webcast. This provides a window through which the public can monitor developments and offer comment; additionally, by advising NIH directors and holding occasional safety and ethics symposia, the RAC also provides a way of addressing scientific, ethical, and/or safety concerns as they emerge.†††

The geography of gene transfer

Gene transfer originated in the USA; it accounts for two-thirds of all phase 1 or 1–2 studies performed worldwide. By 1992, European countries had begun registering gene-transfer trials and, along with the UK, they account for a quarter of early-phase studies.

Countries like Japan, Australia, Israel, and China also established gene-transfer programs early on. As has happened with elite clinical practice and clinical trials more generally, gene

§§ Detailed accounts of the history of RAC review can be found in Wright[61] and Rainsbury.[62]

¶¶ Other countries have different systems for assuring biosafety. In many European countries, gene-transfer studies must be approved by public health and environment authorities. In the UK gene-transfer researchers self-regulate.[63]

*** In the UK the centralized committee is GTAC;[67] in the Netherlands, research ethics review in specialized areas like gene transfer (others are xenotransplantation, cell transplantation, and heroin addiction) is centralized with the CCMO.[69,70]

††† Past symposia have discussed the safety of AAV vectors, cardiovascular gene transfer, lentivirus biosafety, and retroviral gene transfer safety.

transfer has undergone a sort of geographical dispersion over the last few decades. Several of the field's most noteworthy successes in terms of clinical outcomes have occurred on European soil, including three studies involving primary immunodeficiencies. A number of Asian countries – in particular China – have significantly enlarged their gene-transfer research programs. And countries like Brazil,[71] Korea, and the Philippines now have indigenous gene-transfer studies.

At least in the USA, the vast majority of studies are performed at academic medical centers (one FDA official told me that only about 5 percent of gene-transfer studies are not pursued at academic medical centers and do not submit protocols for review to the RAC, though others put the figure near 10 percent).[72,73] Consistent with the geographical dispersion described above, gene transfer has also been pursued in newer settings. Once exclusively the province of major gene transfer research centers like the NIH, University of Pennsylvania, Cornell-Weill Medical Center, or Baylor College of Medicine, some early-phase studies are conducted in private, outpatient clinics, where volunteers might receive study agents from their personal physicians.

The program

To what ends is gene transfer aimed? There are, of course, as many "programs" and ends as there are "actors." Many ethicists and policy makers divide gene transfer into two categories. Therapeutic gene transfer involves correcting disease; examples include gene transfer for treating cancer or cystic fibrosis. Enhancement gene transfer aims at improving nonpathological states, interventions aimed at quality of life, or interventions that add new traits. A third category – investigational gene transfer – is surely appropriate as well. This would include studies involving gene marking and healthy volunteers.

There is considerable fuzziness in the boundary between therapy and enhancement, with the former generally occupying less morally contested terrain (as a consequence, a common rhetorical strategy for investigators pursuing borderline cases is to argue that a condition fits squarely within the therapeutic category). Most credible ethics accounts question whether a moral distinction between enhancement and therapy is apt, noting that well-established practices like childhood vaccination and higher education represent a kind of human enhancement, while "therapeutic" interventions like cochlear implants or heroic, life-extending interventions provide examples of contested therapy.

Nevertheless, many jurisdictions, such as the UK, initially restricted somatic gene transfer to lethal and/or seriously disabling conditions. Over time, these limits have been relaxed, and protocols extended to conditions like hemophilia (which involves effective, if imperfect, standard of care in high-income countries) and later, to healthy volunteers.[74] The first "quality-of-life" protocol that raised serious questions about acceptable risk involved a rare and often hereditary eye cancer, retinoblastoma. The study was aimed at developing an eye-sparing approach. According to news reports, the protocol received a particularly skeptical hearing before the RAC.[72] Nevertheless, a less aggressive version proceeded, without incident.[29] Since then, a number of protocols have been pursued – with ostensible approval from the RAC as well as the public – that fit securely under the heading of lifestyle and enhancement. These include protocols involving control of dental caries, erectile dysfunction, and peanut allergy.

When gene-transfer strategies were first discussed in the 1970s and early 1980s, they were conceived in terms of hereditary (in particular, Mendelian) illnesses. From 1990 through to 1995 inclusive, there was one study aimed at a monogenic disease for every three involving cancer. The next six years saw a dramatic shift, as gene-transfer techniques were extended to

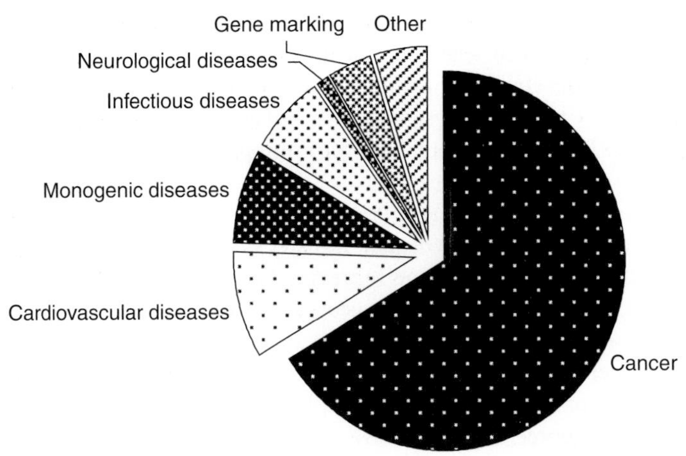

Figure 2.7 Medical indications in gene transfer trials. Derived from Wiley, *Journal of Gene Medicine*, "Gene Therapy Clinical Trials Worldwide" (www.wiley.co.uk/genmed/clinical/; accessed August, 2008).

more prevalent – and potentially more lucrative – indications like limb ischemia, heart disease, and AIDS. In the years 1996 to 2001, the number of genetic-disease gene-transfer studies dropped to one for every 10 cancer studies.[73] Figure 2.7 shows a distribution of different indications tested in gene transfer.

Programmatic orientation of gene-transfer clinical research

Gene transfer is but one manifestation of another, broader shift in health and science policy: translational research. Most accounts of contemporary research policy begin with the extraordinary scientific and technological accomplishments behind the Allied victory in World War II. Innovations like radar, atomic weaponry, mass production of penicillin, and isolation of blood components were made possible because of a novel arrangement in which the US federal government disbursed funding to academia, and left it to networks of scientists to determine precisely how the money was spent.[75]

The architect of this arrangement, Vannevar Bush, envisioned research as a unidirectional process by which fundamental discoveries made in academic laboratories would be taken up by the private sector or government laboratories and transformed into applications. Though Bush was not to have his way in terms of institutional specifics (for example, Bush preferred that funding be administered under a single body rather than being splintered among different agencies), his model of innovation largely prevailed for almost half a century.

The early 1980s saw the first renegotiation of this rigid and hierarchical distinction between the basic and applied. Concerned that overseas firms were free-riding off US discoveries, policy makers passed a series of new laws to encourage rapid commercialization of federally funded academic research. The most famous of these was the University and Small Business Patent Procedures Act (often called the "Bayh-Dole Act"),[76] which allowed universities to seek intellectual property on innovations emerging from federally funded research. Universities responded energetically by seeking patents and creating technology-transfer offices. In 1989, NIH gene-transfer researcher W. French Anderson became one of the earliest beneficiaries of similar legislative reforms when he obtained a Cooperative Research and Development Agreement (CRADA) that transferred the title of NIH-owned patents to his biotechnology firm, Genetic Technologies Inc.[77,78]

A second major shift in research policy began in the first years of the new century. With the NIH budget nearly doubled and the human genome sequence approaching completion, expectation was high that great leaps forward in health care would follow immediately. Instead, productivity in the pharmaceutical industry actually declined,[79] as did the number of new molecular entities and biologics license applications submitted to the FDA.[80] Within a few years, "translational research" became the new mantra for health research policy.[‡‡‡] Longstanding malaise within medicine about the status of and support for clinical research crystallized around reports calling for a translational research career track (see, for example, the AAMC Task Force on Clinical Research[81]). In 2003, NIH director Elias Zerhouni unveiled his "NIH Roadmap," which among other things, included funding for the eventual establishment of 60 Clinical and Translational Science Centers (CTSCs) around the USA. The FDA's Critical Path initiative aimed at identifying scientific developments and regulatory reforms that would hasten the development of new drugs and improve research productivity. And numerous universities broke ground for new clinical research facilities. Similar ventures were undertaken elsewhere, as Europe's drug regulatory agency launched the originally named "Road Map to 2010"[82] (Canadian drug regulators were only slightly more imaginative with their "Blueprint for Renewal"[83]), and foreign research agencies explored the CTSC model.[84]

Though but one facet of a larger enterprise, gene transfer research is in many respects paradigmatic of practices towards which these initiatives aim. For example, stereotypical early-phase gene-transfer studies incorporate numerous basic science discoveries, and provide an occasion for the reciprocal flow of information back to the laboratory. Trials often draw on large, interdisciplinary research teams of the sort envisioned by the Roadmap and AAMC Task Force. Studies often involve collaboration across multiple centers. Many early phase studies in gene transfer are analogous to the types of exploratory, "phase 0" studies encouraged by the FDA. And the difficulties experienced by gene-transfer researchers – prohibitive costs for running toxicology studies, vector production bottlenecks, logistical and funding issues for human trials, regulatory complexity[85–87] – are in many respects typical of those encountered by other translational researchers.

As I write, there is some indication that these initiatives may be bearing fruit, as more drugs enter clinical testing than before.[88] But new political and ethical strains are also apparent. The blurred identities of traditional actors described in the previous section, and the shifts in emphasis and affiliation, create new social, ethical, and political strains: are drug regulators compromising their role as guardians of the public health as they accommodate interest-group pressure and recast themselves as drivers of innovation? Will renewed emphasis on drug development within academia further erode norms of transparency and self-directed inquiry? If so, what new measures or metrics will be introduced to ensure public accountability, and what values will these tacitly normalize?

Conclusion

In Figure 2.8 I provide a timeline of major events over the twenty-year history of gene transfer. It is largely a record of important first accomplishments and major setbacks. It is common for commentators to note the latter, and deride the field of gene transfer for ethical misconduct. Many gene-transfer researchers acknowledge problems in the early years, but nevertheless claim that the field has cleaned up its act and note the great progress made in it. In poring through its research output and media coverage, and conversing with various persons who

‡‡‡ What, precisely, is meant by the term is not clear. In the USA translational research is said to consist of two components: incorporation of laboratory discoveries to clinical practice and vice versa (what is often called "T1"), and application of clinical discoveries to community practice and vice versa (T2).

1949	One of the earliest tests that used viruses to treat human disease (hepatitis virus for Hodgkin's disease).
1975	Stanfield Rogers and colleagues report the first attempted use of viral preparations against a genetic disease (Shope papilloma virus for hyperargininemia).
Jul 1980	UCLA hematologist Martin Cline attempts human gene transfer using recombinant DNA. Cline had not received approval from the Recombinant DNA Advisory Committee (RAC) or the UCLA institutional review board (IRB), and was sanctioned.
Oct 1988	First approved gene-transfer protocol. The trial used a neomycin resistance gene to mark tumor-infiltrating lymphocytes in volunteers with cancer. The same team carried out the first "official" gene-transfer trial to treat a genetic disease, adenosine deaminase deficiency–severe combined immunodeficiency (ADA–SCID).
Dec 1992	First submission of adenoviral vector protocol to the RAC, leading to the first published clinical trial (involving cystic fibrosis).
Sep 1994	First submission of an adeno-associated vector (AAV) protocol to the RAC.
Dec 1995	Arno Motulsky and Stuart Orkin produce an NIH-commissioned report on the status of the field. Their report is highly critical.
Nov 1996	RAC is reorganized with a smaller membership; the committee's approval authority for protocols is eliminated.
Mar 1997	RAC reviews the first-ever gene transfer protocol involving healthy volunteers.
Jul 1997	Jacques Cohen and colleagues report the birth of an infant after using ooplasmic transfer (because ooplasm contains mitochondria, their protocol involved a form of germ cell gene transfer).
Sep 1999	Jesse Gelsinger dies in ornithine transcarbamylase deficiency trial pursued at the University of Pennsylvania, USA. The episode results in a lawsuit that is settled in 2005.
Apr 2000	A team of researchers at Hôpital Necker (Paris) report the first clear demonstration of therapeutic efficacy for gene transfer. The study involves a rare, hereditary immunodeficiency syndrome, X-linked SCID (X-SCID).
Jun 2002	Investigators in the X-SCID study report a lymphoproliferative disorder in one of the volunteers. Four others develop a similar disorder over the next five years, and one dies.
Jun 2002	Milan-based team reports second clear demonstration of therapeutic efficacy of gene transfer, in a trial involving ADA-SCID. With over seven years of follow-up for the first two patients and a median of almost four years for another eight, no lymphoproliferative disorders have been reported.
Oct 2003	China approves the first gene transfer agent (Gendicine) for clinical use, against head and neck cancer.
Apr 2006	A third trial reports promising results against chronic granulomatous disease. However, the clonality of modified cells raises concerns about cancer risk.
Jul 2007	A volunteer dies in arthritis phase 1/2 trial involving AAV vectors. An investigation concludes that a causal role of gene transfer is highly unlikely but cannot be excluded.

Figure 2.8 Timeline of major events in gene transfer.

ally themselves with the field, I have come to the conclusion that "gene transfer" is an elastic, heterogeneous, and evolving category,[§§§] and that one should be very careful when offering generalizations about the "health," character, or integrity of the field or its personnel.

Some describe the field's tumult as emerging from a clash of cultures in which doctors advocate for their patients, and scientists advocate for data.[89] There might be some truth to this, but it ignores rich traditions of caution and evidence evaluation within medicine, as well as the deeply humanitarian motivations of many scientists. Thus, while many will see heroic incaution in the way W. French Anderson, Steven Rosenberg, and Michael Blaese were able

[§§§] Yet one further indication of this elasticity is the recent renaming of gene-transfer professional societies. In 2008, the European Society of Gene Therapy was renamed "European Society of Gene and Cell Therapy"; the US counterpart, the American Society of Gene Therapy, now also goes by the name "American Society of Gene and Cell Therapy."

to overcome public and regulatory skepticism around the first trial, it fell to another set of physicians – Harold Varmus, Arno Motulsky, and Stuart Orkin – to sound the alarm about a rush to clinical testing.

Another common suggestion is that the ethical challenges and travails encountered in translating somatic gene transfer are hardly unique. I agree with this statement in broad outline: financial conflict of interest, uncertainty, unexpected safety problems, therapeutic misconception, hyperbole, and questionable subject selection all occur in other contexts, and a premise of this book is that lessons learned in gene transfer can be applied to other areas of translational research.

But this claim is in tension with two other points. For one, the aspirations, anxieties, public reception, and even the technologies of a field are, to a very large extent, shaped by historical circumstances. Other new platforms – stem cells and nanotechnologies, for example – will encounter a similar configuration of challenges, but under somewhat different circumstances, with different actors, and with different associated social and ethical baggage. Surely, these will affect policy and governance.

Second, there *are* distinctive qualities with gene transfer that I believe herald new phenomena, opportunities, and risks for medicine. I have already noted the field's platform-centered organization, in which many actors within the field appear to affiliate as much with their medical specialty or scientific subdiscipline, as with a technological platform. The many issues catalogued in the previous paragraph cluster together in individual trials, adding to the difficulty in prosecuting studies ethically.

Third, the degree of complexity and indeterminacy of gene transfer, and the lability of the actors and institutions, present new challenges, many of which will be catalogued in this book. Other challenges will likely unfold as gene transfer evolves, and as strategies enter late-stage testing and clinical practice. To the bounded question titling this chapter, part of the response must surely be "we shall see."

References

1. Friedmann T. *Gene Therapy: Fact and Fiction in Biology's New Approaches to Disease*. Cold Spring Harbor, NY, CSH Press, 1983.
2. Anderson WF. Human gene therapy: The initial concepts. In: Brigham KL, ed. *Gene Therapy for Diseases of the Lung*. New York, Marcel Dekker Inc., 1997.
3. Judson HF. The Glimmering Promise of Gene Therapy. *MIT Technology Review* November 1, 2006.
4. Hu J. Gene therapy: Back to basics. In: Ma J, ed. *Gene Expression and Regulation*. Beijing, Springer, 2006.
5. Rogers S. Shope papilloma virus: A passenger in man and its significance to the potential control of the host genome. *Nature* 1966; **212**: 1220–2.
6. Terheggen HG, Schwenk A, Lowenthal A, Van Sande M, Colombo JP. Argininaemia with arginase deficiency. *Lancet* 1969; **294**(7623): 748–9.
7. Terheggen HG, Lowenthal A, Lavinha F, Colombo JP. Familial hyperargininaemia. *Arch Dis Child* 1975; **50**(1):57–62.
8. Terheggen HG, Lowenthal A, Lavinha F, Colombo JP, Rogers S. Unsuccessful trial of gene replacement in arginase deficiency. *Z Kinderheilkd* 1975; **119**(1): 1–3.
9. Leydesdorff L, Etzkowitz H. Emergence of a triple helix of university–industry–government relations. *Science and Public Policy* 1996; **23**(5): 279–86.
10. Jasanoff S. *Designs on Nature: Science and Democracy in Europe and the United States*. Princeton, Princeton University Press, 2005.
11. Jasanoff S. Product, Process, or Programme: Three Cultures and the Regulation of Biotechnology. In: Bauer M, ed. *Resistance to New Technology: Nuclear Power, Information Technology, and Biotechnology*. New York, Cambridge University Press, 1995.
12. Tilghman SM. Address to the Stem Cell Institute New Jersey, November 11, 2004.

13. www.ornl.gov/sci/techresources/Human_Genome/medicine/genetherapy.shtml#whatis.
14. www.asgt.org/about_gene_therapy/defined.php.
15. American Society of Gene Therapy, *Guide for Press Contacts*.
16. *Position paper on social, ethical and public awareness issues in gene therapy*. November 2002. At: www.esgct.org/downloads/position_2.pdf.
17. http://stats.oecd.org/glossary/detail.asp?ID=6306 (accessed June 30, 2008).
18. Krimsky S. *Biotechnics and Society: The Rise of Industrial Genetics*. Westport, CT, Praeger, 1991; 167.
19. Braat H, Rottiers P, Hommes DW, Huyghebaert N, Remaut E, Remon JP, *et al.* A phase I trial with transgenic bacteria expressing interleukin-10 in Crohn's disease. *Clin Gastroenterol Hepatol* 2006; **4**(6): 754–9.
20. Shanahan F. Immunology. Therapeutic manipulation of gut flora. *Science* 2000; **289**(5483): 1311–12.
21. Council of Europe. *Convention for the Protection of Human Rights and Dignity of the Human Being with regard to the Application of Biology and Medicine: Convention on Human Rights and Biomedicine*. Series 164; 1997.
22. 37th Canadian Parliament. An Act Respecting Assisted Human Reproduction and Related Research. *Bill C-6*, 3rd edn, Canada, 2004.
23. Indian Council of Medical Research. *Ethical Guidelines for Biomedical Research on Human Participants*. New Delhi, Royal Offset Printers, 2006.
24. Firfer H. How far will couples go to conceive? CNN.com, 2004.
25. Chapman AR, Frankel MS, eds. *Designing Our Descendants: The Promises and Perils of Genetic Modifications*. Baltimore, Johns Hopkins University Press, 2003.
26. Rasko J, O'Sullivan G, Ankeny R, eds. *The Ethics of Inheritable Genetic Modification: A Dividing Line?* Cambridge, Cambridge University Press, 2006.
27. Wu Z, Asokan A, Samulski RJ. Adeno-associated virus serotypes: vector toolkit for human gene therapy. *Mol Ther* 2006; **14**(3): 316–27.
28. Recombinant DNA Advisory Committee (RAC). Minutes of Meeting June 14, 1999.
29. Chevez-Barrios P, Chintagumpala M, Mieler W, Paysse E, Boniuk M, Kozinetz C, *et al.* Response of retinoblastoma with vitreous tumor seeding to adenovirus-mediated delivery of thymidine kinase followed by ganciclovir. *J Clin Oncol* 2005; **23**(31): 7927–35.
30. John Wiley and Sons. Gene Therapy Clinical Trials Worldwide Database. 16 August, 2008.
31. Manno CS, Pierce GF, Arruda VR, Glader B, Ragni M, Rasko JJ, *et al.* Successful transduction of liver in hemophilia by AAV-Factor IX and limitations imposed by the host immune response. *Nat Med* 2006; **12**(3): 342–7.
32. Kaiser J. Biomedical research. A cure for medicine's ailments? *Science* 2006; **311**(5769): 1852–4.
33. Kimmelman J. Valuing risk: the ethical review of clinical trial safety. *Kennedy Inst Ethics J* 2004; **14**: 369–93.
34. National Institutes of Health. *Estimates of Funding for Various Diseases, Conditions, Research Areas*. Department of Health and Human Services, 2008.
35. Search of Cancer Research Portfolio. National Cancer Institute. http://researchportfolio.cancer.gov/. Accessed June 24, 2008.
36. Kaiser J. The winners. Even in a time of plenty, some do better than others. *Science* 2001; **292**(5524): 1995–7.
37. Crofts C, Krimsky S. Emergence of a scientific and commercial research and development infrastructure for human gene therapy. *Hum Gene Ther* 2005; **16**(2): 169–77.
38. Riordan T. A biotech company is granted broad patent and stock jumps. *New York Times* 1995 March 22.
39. Gelsinger vs. Trustees of the University of Pennsylvania. In: Sherman, Silverstein, Kohl, Rose and Podolsky, Law Offices, Philadelphia, 2000.
40. Recombinant DNA Advisory Committee (RAC). Protocol List. www4.od.nih.gov/oba/rac/PROTOCOL.pdf; 2008.
41. http://clinicaltrials.gov (searched June 24, 2008)
42. University of Pennsylvania Medical Center Announces First Human Gene Therapy Institute in the World. *PR Newswire* 1992 December 7.

43. Nathan's Story. www.nathansbattle.com/meetnathan/story.html. Accessed June 25, 2008. Ironically, the Batten study did not receive full RAC review.
44. Goss CH, Mayer-Hamblett N, Kronmal RA, Ramsey BW. The cystic fibrosis therapeutics development network (CF TDN): a paradigm of a clinical trials network for genetic and orphan diseases. *Adv Drug Deliv Rev* 2002; **54**(11): 1505–28.
45. Lara PN Jr, Paterniti DA, Chiechi C, Turrell C, Morain C, Horan N, *et al.* Evaluation of factors affecting awareness of and willingness to participate in cancer clinical trials. *J Clin Oncol* 2005; **23**(36): 9282–9.
46. Seidenfeld J, Horstmann E, Emanuel EJ, Grady C. Participants in phase 1 oncology research trials: are they vulnerable? *Arch Intern Med* 2008; **168**(1): 16–20.
47. Ho J, Pond GR, Newman C, Maclean M, Chen EX, Oza AM, *et al.* Barriers in phase I cancer clinical trials referrals and enrollment: five-year experience at the Princess Margaret Hospital. *BMC Cancer* 2006; **6**: 263.
48. Aldhous P. Interview: Clinical trials of Life. *New Scientist* 2007; **195**(2623): 48–9.
49. Pattee SR. Protections for participants in gene therapy trials: a patient's perspective. *Hum Gene Ther* 2008; **19**(1): 9–10.
50. Nih Office of Biotechnology Activities. *NIH Guidance on Informed Consent for Gene Transfer Research*, 2003.
51. Evans JH. *Playing God?: Human Genetic Engineering and the Rationalization of Public Bioethical Debate*. Chicago, University of Chicago Press, 2002.
52. Anderson WF. Musings on the struggle – Part III: The October 3 RAC meeting. *Hum Gene Ther* 1993; **4**(4): 401–2.
53. Alper J. Biotech thinking comes to academic medical centers. *Science* 2003; **299**: 1303–5.
54. Couzin J. Clinical research. Advocating, the clinical way. *Science* 2005; **308**(5724): 940–2.
55. Moukheiber Z. Drug Money. *Forbes.com* 2001.
56. Groopman J. Buying a cure. *New Yorker – Online* 2008 January 8.
57. Kessler DA, Siegel JP, Noguchi PD, Zoon KC, Feiden KL, Woodcock J. Regulation of somatic-cell therapy and gene therapy by the food and drug administration. *N Engl J Med* 1993; **329**: 1169–73.
58. Korwek EL. What are biologics? A comparative legislative, regulatory and scientific analysis. *Food Drug Law J* 2007; **62**: 257–304.
59. Noguchi PD. From Jim to gene and beyond: an odyssey of biologics regulation. *Food Drug Law J* 1996; **51**(3): 367–73.
60. See, for example, International Conference on Harmonisation of Technical Requirements for Registration of Pharmaceuticals for Human Use (ICH) (2002), *Communication Paper: The First Workshop on Gene Therapy*. Available for download at: www.ich.org/cache/compo/276–254–1.html. Accessed June 27, 2008.
61. Wright, S. *Molecular Politics: Developing American and British Regulatory Policy for Genetic Engineering 1972–1982.* Chicago, University of Chicago Press, 1994.
62. Rainsbury JM. Biotechnology on the RAC – FDA/NIH regulation of human gene therapy. *Food Drug Law J* 2000: **55**: 575–600.
63. Deans J, *et al.* (2005) Gene Therapy – Potential Health and Safety Implications. (Horizon Scanning Intelligence Group, ed), www.hse.gov.uk/horizons/genetherapy/hsigjuly054.pdf (accessed June 26, 2008).
64. National Institutes of Health. *NIH Guidelines for Research Involving Recombinant DNA Molecules* (NIH Guidelines), 2002, at: IV-B-2-a-(7).
65. Race MS, Hammond E. An evaluation of the role and effectiveness of institutional biosafety committees in providing oversight and security at biocontainment laboratories. *Biosecur Bioterror* 2008; **6**(1): 19–35.
66. Fontes B, Agentis T. Laboratory-acquired vaccinia infection at a small research institution. *ABSA Conference*, Nashville, TN, 2007.
67. www.advisorybodies.doh.gov.uk/genetics/gtac
68. Trent RJ. Oversight and monitoring of clinical research with gene therapy in Australia. *Med J Aust* 2005; **182**(9): 441–2.
69. Bleijs DA, Haenen IT, Bergmans JE. Gene therapy legislation in The Netherlands. *J Gene Med* 2007; **9**: 904–909.
70. www.ccmo-online.nl

71. Michaluart P, Abdallah KA, Lima FD, Smith R, Moyses RA, Coelho V, *et al.* Phase I trial of DNA-hsp65 immunotherapy for advanced squamous cell carcinoma of the head and neck. *Cancer Gene Ther* 2008; **15**(10): 676–84.
72. Garber K. RAC nixes plan to treat retinoblastoma. *Science* 1999; **284**(5423): 2066.
73. Based on search of Wiley database, June 19, 2008.
74. Marshall E. Panel approves gene trial for "normals". *Science* 1997; **275**(5306): 1561.
75. Kleinman DL. *Politics on the Endless Frontier: Postwar Research Policy in the United States*, Chapter 4. Durham, Duke University Press, 1995; 60.
76. 35 USC § 200–212: Patent Rights in Inventions Made Under Federal Funding Agreements.
77. Culliton BJ. NIH, Inc.: the CRADA boom. *Science* 1989; **245**(4922): 1034–6.
78. Henig RM. Dr. Anderson's Gene Machine. *New York Times* 1991 March 31.
79. Service RF. Surviving the blockbuster syndrome. *Science* 2004; **303**(5665): 1796–9.
80. US Food, and Drug Administration. *Innovation or Stagnation? Challenge and Opportunity on the Critical Path to New Medical Products*. US Department of Health and Human Services, 2004.
81. Association of American Medical Colleges. *Promoting Translational and Clinical Science: The Critical Role of Medical Schools and Teaching* Hospitals (Report of the AAMC's Task Force II on Clinical Research), 2006 May.
82. Milne C.-P. US and European regulatory initiatives to improve R&D performance. *Expert Opinion in Drug Discovery* 2006; **1**(1): 11–14.
83. Ministry of Health. *Blueprint for Renewal II: Modernizing Canada's Regulatory System for Health Products and Food*. Health Canada, Ottawa, 2007.
84. Butler D. Translational research: crossing the valley of death. *Nature* 2008; **453**(7197): 840–2.
85. Woo SL, Skarlatos SI, Joyce MM, Croxton TL, Qasba P. Critical resources for gene therapy in Heart, Lung, and Blood Diseases Working Group. *Mol Ther* 2006; **13**(4): 641–3.
86. Nyberg KA, High KA, Salomon DR. *Challenges in Advancing the Field of Gene Therapy: A Critical Review of the Science, Medicine, and Regulation*. American Society of Gene Therapy, Arlington, VA, April 7–8 2005.
87. Friedmann T. Changing roles for academia and industry in genetics and gene therapy. *Mol Ther* 2000; **1**(1): 9–11.
88. *Outlook 2007*. Boston, MA, Tufts Center for the Study of Drug Development, 2007.
89. Angier N. Cultures in conflict: M.D.'s and Ph.D.'s. *The New York Times* 1990 Apr 24.

Chapter 3

Safety, values, and legitimacy: the protean nature of risk in translational trials

Introduction

Monday morning, September 13, 1999, a team of physicians led by Steven Raper and James Wilson at the University of Pennsylvania threaded a thin tube through an incision in the groin of human subject OTC.019 to administer thirty milliliters of a modified adenovirus to his liver. The subject, like the seventeen who had preceded him, suffered from a rare hereditary disorder, ornithine transcarbamylase deficiency (OTCD), in which patients accumulate toxic levels of ammonia. Half of all infants born with severe forms of the disorder die within seventy-two hours of birth. However, participants in this study suffered from a mild form – one that could be controlled with medication and a restricted protein diet.

Like the previous subjects, OTC.019 developed the usual symptoms of viral infection: achiness, fevers, and a headache. Eighteen hours after the infusion, however, the response of OTC.019 headed into unfamiliar territory. He began showing signs of jaundice, a troubling sign for someone with a liver disorder. He became disoriented and agitated. Tests indicated dangerous levels of ammonia in the volunteer's blood, and soon thereafter he slipped into a coma.

Thirty or so hours after receiving the vector, OTC.019 began hyperventilating. Elevated oxygen threatened to increase the rate at which his body broke down the proteins released by his deteriorating blood cells; this, in turn, would drive his ammonia levels higher, portending brain damage. To control his breathing and reduce his blood oxygen, physicians placed OTC.019 on a ventilator. But OTC.019 continued to hyperventilate. The medical team induced a deeper coma. Now, the opposite problem: blood oxygen levels began an inexorable decline. At eighty-eight hours, the liver and kidneys of OTC.019 began failing. Ninety-eight hours after having delivered the vector, physicians withdrew life support, and OTC.019 expired.

Two weeks later, OTC.019 was publicly identified as Jesse Gelsinger, and his death quickly came to symbolize all that is wrong with gene transfer, research oversight, ethicists, academia, and drug regulation. The episode has since become a sort of moral Rorschach test. The anthropologist Mary Douglas once described risk as a "forensic resource" – that is, a device that social groups use for assigning (or deflecting) blame and moral responsibility.[1] Gelsinger's death is one such forensic resource, with some viewing the episode as a show of how money, ambition, and science can combine to make a deadly mix, others seeing a tragic but unavoidable outcome with cutting-edge research, and still others considering this as a case of bad apples. In what follows, I examine various interpretations of why Gelsinger died, and explore what these interpretations say about how risks should be assessed and disclosed in translational research trials.

Adenoviral vectors, non-linear dose–toxicity curves, and innate immunity

The standard scientific account is that adenoviral vectors killed Gelsinger by inducing a "cytokine cascade that led to disseminated intravascular coagulation, acute respiratory distress, and multi-organ failure."[2,3] Some have suggested that the event was "unexpected."[4–7]

Clearly, such an explanation is incomplete. For one thing, though Gelsinger's death was undoubtedly accidental, whether it was unexpected is disputed. Adenovirus vectors had long been recognized as capable of provoking powerful immune reactions. In the first ever adenoviral gene-transfer studies in humans, a cystic fibrosis patient developed lung inflammation at vector doses significantly lower than those predicted from animal studies.[8] An additional concern for adenoviruses was the "biphasic" shape of their dose–toxicity curve. In preclinical studies, researchers found that animals showed only moderate signs of toxicity at low doses. As they gradually escalated vector levels, toxicity increased precipitously. This biphasic dose–toxicity curve was particularly worrisome for researchers, because they could never be sure whether they were administering doses on the flat (and safe) portion of the curve, or its steep (and noxious) tail.

Both issues were discussed when Raper and Wilson presented their protocol to the Recombinant DNA Advisory Committee (RAC), a safety and ethics review panel located within the National Institutes of Health (NIH).[9] Panelist Robert Erickson, a pediatric geneticist, voiced concerns about the vector triggering liver inflammation, thereby plunging otherwise asymptomatic subjects with a liver disorder into medical crisis. Erickson also worried about the biphasic toxicity curve in the preclinical studies. Perhaps even more prescient were the remarks of NIH virologist Stephen Straus, who questioned whether safety studies performed in non-human primates were useful for predicting human response, given that primates resist adenovirus infection. Straus knew a thing or two about the perils of projecting human toxicity from animals. Two years earlier, he had collaborated in a trial investigating the effects of fialuridine (FIAU) for treating hepatitis B. The drug had an unexpectedly long half-life in humans, and five of twenty-four participants died from liver toxicity. The deaths prompted a series of investigations and lawsuits.[10,11]

The decision to enroll medically stable volunteers: ethics and eligibility

Neither scientists, ethicists, Gelsinger's father, regulators, nor members of the public appear content to blame (amoral) viruses. In his suit against the University of Pennslvania, Paul Gelsinger (Jesse's father) named Arthur Caplan – perhaps the USA's most publicly visible bioethicist – as a defendant. Caplan was instrumental in persuading the research team to recruit asymptomatic adults like Gelsinger.

Investigators might instead have conducted the study in terminal infants. Most phase 1 oncology trials, as well as studies of gene transfer, are performed in volunteers whose disease is advanced and refractory to standard treatments. This is sometimes referred to as the "oncology model" of subject selection, because oncology phase 1 studies almost universally enroll such patients. The oncology model offers subjects the most favorable balance of risks and benefits. But it also presents challenges for the exercise of autonomy, and Caplan argued that desperate parents of dying infants would be unable to provide valid informed consent. In addition, underlying morbidities of treatment-refractory patients often confound causal attribution when adverse events occur. For scientific reasons as well, the OTCD study team preferred to avoid infants in medical crisis.*

* The researchers proposed performing liver biopsies on patients in order to test gene expression. This would not have been permissible in minors under US regulations because biopsy risks would be greater than minimal and would lack therapeutic justification.

Instead, Caplan advocated what might be termed the "stable volunteer" model of subject selection, in which volunteers are disease-afflicted, but their symptoms are medically controlled. Like the healthy volunteer studies (a third model of subject selection), the stable volunteer model offers favorable circumstances for informed consent. Yet because they threaten the adequate medical status of volunteers for only incremental medical gain, these studies involve a less favorable risk–benefit balance for volunteers. In short, the choice to pursue the OTCD study in medically stable subjects involved prioritizing informed consent and scientific value over medical beneficence.

The selection of "stable volunteers" has frequently been condemned in the trial's aftermath. But this interpretation of the OTC debacle also proves unsatisfactory. If the ethicality of recruiting stable volunteers to the study was patent, why did an initially dubious RAC ultimately come around to the study's subject-selection criteria? And it should be noted that RAC's membership included several ethicists, as well as an outspoken leader of a patient-advocacy organization. The use of stable volunteers was also endorsed by the National Urea Cycles Disorder Foundation (NUCDF), which represents people with OTCD. For that matter, gene transfer's very first "therapeutic" (as opposed to marking) trial involved subjects whose condition, ADA-SCID, was medically controlled. Stable volunteers continue to be recruited into various other gene-transfer trials, including a subsequent one involving intravenous administration of modified adenovirus vectors.*

Violations of study protocol

Then there is the "good investigators gone bad" explanation. When Raper and Wilson submitted their trial proposal to the RAC, they proposed numerous design features – eligibility criteria, stopping rules, etc. – to enhance the safety of their study. Following Gelsinger's death, an FDA audit revealed that investigators strayed significantly from the protocol. For example, their protocol excluded many categories of subjects who might be susceptible to the vector's toxicity. In numerous instances, the investigators fudged these criteria. Thus, Gelsinger's blood ammonia levels, though falling within the protocol's inclusion criteria at the time of screening, exceeded permissible levels at the time of dosing.[14] Similarly, the investigators (in some cases without permission from the FDA) enrolled several subjects whose ammonia levels exceeded exclusion levels.

Another design feature in the approved protocol was that two female subjects would be enrolled in each dose cohort before dosing a male subject. The rationale was that, because female subjects were heterozygous for the liver disorder, they would be less susceptible to liver toxicity. The investigators seem to have hewed to this restriction until Gelsinger, who was enrolled as the second subject in his cohort. To make sure that toxic responses to the vector subsided, investigators also planned to wait thirty days before administering the vector to a new subject. In one instance, however, the researchers waited only fourteen.

The protocol was to be halted the moment significant (i.e. grade 3 or 4) toxicities emerged in any subject. Instead, investigators sought permission from the FDA to continue each time significant toxicities were observed. On several occasions the investigators seem to have forged ahead without FDA approval.[15†] Similarly, investigators paid little heed to the requirement

---

* This refers to Genstar's study using highly deleted adenoviral vectors against hemophilia. The study was halted after the first subject experienced a transient rise in serum transaminases and thrombocytopenia.[12,13]

† For example, FDA's Thomas Eggerman testified (from minutes): "If two patients developed grade 2 NCI Common Toxicity Criteria toxicities or if a single patient developed a grade 3 or higher toxicity, the study would be put on clinical hold pending an explication acceptable to the University of Pennsylvania Institutional Review Board (IRB) and to the FDA .... Cohort 4 (dose level, 6 x 109 particles/kg) included

that the study be stopped whenever two mild (grade 1 or 2) toxicities were observed within a cohort.

Finally, concurrently with their OTCD study, Wilson and Raper conducted a series of experiments testing the safety of hepatic administration of first-generation adenoviral vectors in rhesus monkeys. Two monkeys were euthanized because of toxicity within five days of receiving vector at doses comparable to that intended in the human study. According to the FDA, these toxicities were not immediately disclosed to the agency, nor to subjects.[15] They had, however, been submitted to a journal seven months before Gelsinger received vector, and were published a month after he died.[17]

These and other protocol deviations leave the impression of a rushed and cavalier approach. A cornerstone of modern research ethics is the independent assessment of a study's risks. In the multiple lapses in reporting protocol modifications, the investigators clearly acted unethically.

But before rushing to judge the investigators' conduct, three points should be considered. First, many protocol variations appear to have enjoyed the assent of the FDA; in reading through warning letters to the investigators, one cannot help but form the impression that the agency was straining to exculpate itself. Second, there seems to be little evidence to suggest that protocol deviations, like Gelsinger's ammonia levels or his inclusion as the second subject in his cohort, triggered the fatal immune response. Third, there is only limited evidence on which to conclude that protocol deviations in this study were more frequent or extreme than in other studies.[‡] Eligible subjects are often difficult to find for rare disorders like mild OTCD. Perhaps even harder to find are eligible subjects willing to receive a genetically engineered virus and take a needle to their bellies for a liver biopsy. Furthermore, trials often necessitate major logistical disruptions for participants. When a "consented" subject flies in from Arizona, and one of his laboratory parameters falls just outside the inclusion window, it's not difficult to understand how investigators might be tempted to stretch their criteria.

Personal ambition and investigator misconduct

From where did this cavalier approach originate? By all accounts, James Wilson was, and remains, an extraordinarily ambitious researcher. Wilson began his scientific career at the University of Michigan where, with his mentor William Kelley, he cloned several gene variants involved in Lesch–Nyhan disease. In 1989, Kelley moved to the University of Pennsylvania to take a post as Dean of the School of Medicine.[19] Under his leadership the University of Pennsylvania School of Medicine moved from 10th to 2nd in NIH funding.[19] Kelley recruited Wilson to Penn in 1993 to direct a new program on human gene therapy.[20] Wilson

---

one patient with transient liver enzyme elevations rated at grade 3, which triggered the stopping rule. After receiving approval from the FDA to enroll a third patient, that patient also developed transient liver enzyme elevations rated at grade 3. Although the protocol dictated that the stopping rule had been met, the sponsor did not contact the FDA to review these results before treating a fourth patient in the cohort. The same results occurred for the fourth patient, and the sponsor submitted summary data for the third and fourth patients approximately 2 months after their treatments. However, this result was not discussed with the FDA at the end of cohort 4 (prior to beginning cohort 5), an aberration from the sponsor's actions after cohorts 1, 2, 3, and 5, which featured discussions with the FDA prior to progressing to the next cohort."[16]

‡ In the aftermath of Gelsinger's death, the FDA audited centers where gene transfer trials were being conducted. Of their sample, 47 percent of the research sites had committed minor violations for which voluntary corrective action was urged. Three of the 59 sites audited required official corrective action

quickly became a leader in human gene transfer, spearheading trials for cystic fibrosis,[21] hypercholesterolemia,[22] and neurooncology.[23] Years after the OTCD study, the Institute of Scientific Information placed Wilson on its "highly cited researcher" list – a distinction that indicated his exceptional scientific productivity.

Wilson would appear to be the archetypal "clinical champion:" the academic researcher who doggedly pursues the application of basic science discoveries to a particular disease.[24] Clinical champions are thought by many to be a critical ingredient in medical innovation, and have been widely celebrated in the medical literature.[25,26] Yet they present a major challenge to human protection. Among the traits of the clinical champion is a conviction that agents considered too toxic by others will prove otherwise;[24] these champions are often willing to take risks that depart from consensus opinion.

Wilson nearly said as much in an interview after Gelsinger's death. "It's about leadership and notoriety and accomplishment. Publishing in first-rate journals. That's what turns us on. You've got to be on the cutting edge and take risks if you're going to stay on top."[27] Clinical champions are common in fields like gene transfer, though many researchers have sought to distance themselves from the "cowboy" stereotype. Other examples include Steven Rosenberg (whose trials were once publicly criticized by the NIH for outpacing their evidence base),[28] Ronald Crystal (the first to pursue gene transfer studies in healthy persons),[29,30] and Richard Hurwitz (who pursued a controversial trial using adenoviral vectors to treat a rare, pediatric eye cancer. Standard care would have involved removing the tumor-bearing eye; the protocol aimed at sparing the patients' eye, but at the risk of introducing tumor cells into the patients' circulation).[31] The attraction these clinical champions have for risk is not necessarily a moral deficiency (their perseverance is something we should celebrate), nor am I suggesting that such researchers have no regard for the welfare of their subjects. Rather, fields involving highly lethal and morbid diseases, non-validated therapeutic platforms, large development and production costs, and highly uncertain funding tend not to attract the risk averse.

And yet, Wilson is too easy a target. Blaming him for Gelsinger's death would seem to fulfill a yearning for colorful villains who can be surgically removed from the scene of the crime. Such explanations satisfy "a penchant for psychological explanations, an inability to identify the structural and cultural causes, and a need for a straightforward, simple answer that can be quickly grasped."[32] Moreover, trials "take a village", enlisting the support of other physicians, many of whom feel more at home seeing patients than collecting data. Mark Batshaw – a pediatrician and the third clinical investigator involved in the study – was and remains a prolific researcher. But he also wrote multiple books bearing titles like *When Your Child has a Disability*, which is directed at families, educators, social workers, and advocates (among others).[33] Cowboy-researchers tend not to invest their energies like that. It is partly up to such collaborators, and others like the IRBs, department chairs, and funders, to channel vaulting ambition to constructive ends.

Financial conflict of interest

In an essay written after his son's death, Paul Gelsinger wrote "what is wrong is that a growing, ambitious minority of researchers and institutions have compromised their ethics for profits and prestige, mostly as a result of industry's inappropriate financial influence over them and

(involving, for example, a warning letter). The figures for gene transfer were consistent with those for other, non-gene transfer clinical trials, and were generally better than those for phase 3 trials. Non-adherence to the protocol – in particular, violating entry criteria – is among the most common violations.[18]

our government."[34] His attorney, Alan Milstein, described the field as plagued by "NASDAQ medicine."[35] Perhaps no other issue has garnered greater attention than the OTCD's study's web of financial interests.

Commercial relationships in the study were complex and multiple. They begin with James Wilson's invention of a series of gene-transfer technologies, including the intervention used in the OTCD study, for which he sought patent protection in 1997.[36] As an inventor, Wilson was entitled to a portion of fees and royalties collected by the University of Pennsylvania on the technique. According to the complaint submitted by Gelsinger's attorneys, the biotechnology company Genovo licensed Wilson's liver gene-transfer technology and several others, and it granted Penn the option to negotiate exclusive licenses on techniques later developed at the Institute for Human Gene Therapy (IHGT).[37] Presumably, then, Wilson was receiving direct compensation from Genovo during the study.

As a co-founder of Genovo, Wilson also held up to 30 percent of the company's stock.[38] Under the conflict of interest policy of the University of Pennsylvania, equity interests above 15 percent should have disqualified participation in the study. However, news reports allege that the university altered its policy to enable Wilson's involvement[39] (Wilson, by the way, cashed out his equity at $13.5 million when Genovo was sold to Targeted Genetics in 2000).[40] Meantime, the University of Pennsylvania held 5 percent of Genovo's equity interest, and Genovo agreed to provide the IHGT a budget of $4 million – approximately a fifth of its budget – for 5 years.[38,41] Finally, William Kelley, who as dean had some control over human research policies at the University of Pennsylvania, was an inventor whose name was on several gene-transfer patents, one of which covered all *ex vivo* techniques.[42,43]

Gelsinger's death came to be seen as a sort of "sentinel case" exposing the hazards of financial conflicts in research. In the study's aftermath, a number of institutions and societies, including the American Society of Gene Therapy, issued policies on individual and/or institutional conflict of interest in medical research.[44–47] But three major provisos should be borne in mind when blaming financial interests. First, the study was sponsored by the NIH; any support the study received from Genovo was thus an indirect product of its sponsorship of the IHGT. Second, though Wilson was named as a principal investigator on the study, Raper was the main team member charged with Gelsinger's care, and no one has alleged that he had any financial interests. Finally, as one ethicist put it, "there is almost never a smoking gun [showing adverse outcomes are due to conflicting financial interests]. You can't say Jesse Gelsinger died because Jim Wilson had stock in Genovo."[37]

Problems with informed consent

Another prominent theme in Paul Gelsinger's criticisms was the quality of consent. Testifying before the Senate Subcommittee on Health, Education, Labor, and Pensions, Gelsinger claimed that investigators had led him and his son to believe that gene-transfer research was in a more advanced stage of development than was the case, and that the researchers withheld information on the consent document about the experiments showing the vector's toxicity in rhesus monkeys. He also stated that researchers failed to describe the toxicities they had observed in the study's previous subjects.[48] Nor did the consent document disclose various individual or institutional financial interests.[§] Count four of Gelsinger's suit against the University of

[§] This is based on a copy of the informed consent document received from the OBA under a Freedom of Information Act request. There continue to be allegations that the actual consent document used in the study was different from that submitted to the OBA.

Pennsylvania alleged at least six disclosure failures.[38] Other commentators have expanded on these criticisms. In an introductory bioethics textbook, one writer used the episode in a chapter titled "Research Ethics and Informed Consent."[49]

Unlike many other instances in early-phase gene transfer (discussed in Chapter 4), Gelsinger was told that participation would not provide therapeutic benefits. Paul Gelsinger nevertheless alleged that "the risks were downplayed."[50] To researchers aggrieved by the seeming lengthening of consent documents, Gelsinger's view might seem tainted by hindsight bias. But the suggestion is a plausible one. My own analysis of phase 1 gene-transfer consent documents between 1989 and 2002 found that as many as 20 percent of statements within risk sections moderate or diminish disclosed harms by describing actions investigators will take to reduce harms (e.g. analgesics will be given) or reasons why a risk should not cause alarm (e.g. however, swelling will subside on its own).[ʖ]

How convincing is the case a bad consent process played a role in Gelsinger's death? Might disclosing prior adverse events, including laboratory studies, have given Gelsinger pause? There is no way of knowing; surprisingly, empirical study of how physical risk disclosures influence consent decisions is very limited. Would disclosure of financial interests have deterred Gelsinger? Data are mixed on the degree to which such disclosures matter to typical volunteers.[51–54] Would a consent process better aligned with prevailing ethical and legal norms have provided the Gelsingers with the information they needed to render an informed decision? Probably. But patent inventorship, deceased rhesus monkeys, and fevers have little bearing on what the Gelsingers really needed, which was information on whether the University of Pennsylvania research team, and the various institutions arrayed around the study, could be trusted to protect and advance their interests.

Nevertheless, for reasons that I believe are less ethical than legal psychological, and anthropological, the consent process is one of the most frequent targets for criticism whenever people get hurt in clinical research. In July 2007, 36-year-old Jolee Mohr became the second volunteer with a non-lethal condition to die while participating in a gene-transfer study. The trial was a phase 2 study testing AAV vectors for the treatment of rheumatoid arthritis; Mohr died from a massive fungal infection. Within two months, a review by the RAC all but ruled out the vector as the cause of death.[55] Yet in the aftermath, the episode came to be viewed as a lesson in informed consent. The *Washington Post* ran a highly critical story describing how Mohr had been recruited by her personal physician, and not given sufficient time to reflect on the solicitation.[56] The American Society of Gene Therapy dedicated its annual ethics session to the practice of informed consent.[57] And none other than James Wilson, as editor of the journal *Human Gene Therapy*, used Mohr's death as an occasion to commission a series of essays on informed consent.[58–60]

System failures: FDA, RAC, and IRBs

Many have also suggested that Gelsinger's death was the result of various deficiencies in the system of human subjects protections. Around the time the Penn team initiated their study, the US government began issuing a series of reports warning that human subject protections had not kept pace with the rising volume of clinical research. The reports suggested that IRBs were overburdened, and had insufficient expertise to review many studies.[61,62] Another major concern identified in the reports was the inherent conflict of local ethical review. On the one

ʖ Consent documents were chosen at random, and analyzed by two coders. Reliability was high (exceeding a Cohen's kappa of 0.80). Full data on file with author.

hand, IRBs have a clear mandate to protect the interests of research subjects. On the other, medical centers derive significant status and financial benefits from hosting clinical trials, especially where translational research is involved.[63] In the case of the OTC study, Arthur Caplan reported directly to James Wilson – a relationship that an independent panel that reviewed the IHGT called "unwise."[64] In other realms of risk management, self-regulation has had a spotty record of success;[39] the government's reports questioned whether IRBs were negotiating the conflict adequately.[61,62]

In 1997, the death of a healthy volunteer at the University of Rochester reinforced a perception that protections were failing,[65] as did the shutdown of research at several prominent research centers.[66,67] Just months prior to Gelsinger's death, an NIH panel recommended that the institute's Office for the Protection from Research Risks be strengthened and moved out of the NIH and into the HHS, where it would have greater authority. The panel's recommendations were promptly implemented; upon taking directorship of the new Office of Human Research Protections, one of Greg Koski's first challenges was confronting the political fallout of the OTC study. In the words of the secretary of Health and Human Services, "[t]he tragic death of Jesse Gelsinger focused national attention on the inadequacies in the current system of protections for human research subjects."[68]

Another system failure was the relationship between the two bodies charged with overseeing the safety and ethics of gene-transfer trials. The RAC was originally established to address safety and ethical concerns surrounding gene-splicing research; following revelations that Martin Cline had introduced a recombinant DNA product into human volunteers, the RAC established a Gene Therapy Working Group to review human gene-transfer studies. By 1985 the working group had issued guidelines requiring that investigators working at NIH-funded institutions submit protocols to the RAC for approval. The FDA similarly asserted its authority to regulate gene transfer in 1985.[69]

Regulatory agencies prefer not to see their decisions second-guessed by other authorities. Whether because of rivalry or inactive lines of communication, the RAC was never notified of the decision to change the administration route from intravenous (in the original proposal) to hepatic (in the protocol).[70] Nor was it kept informed about the number of adverse events that had been reported in previous adenovirus trials. An investigation following Gelsinger's death pointed up another seeming weakness in the system. First, only 37 of 970 serious adverse events involving adenoviruses had been reported to the NIH, as required under its guidelines. Second, what might explain the low figures? According to some observers, NIH guidelines required reporting of all adverse events, which includes a universe of unrelated incidents like headaches, care accidents, or manifestations of a natural disease course. To these observers, low adverse-event reporting had less to do with researcher recalcitrance than with poorly crafted guidelines. Third (and in spite of the first two), why hadn't the NIH detected the anomalously low figure, and actively sought toxicity reports from investigators?[71] In hearings on Gelsinger's death, former RAC member Leroy Walters testified about a lack of appropriate cooperation between the FDA, RAC, and the NIH, and the failure of NIH to execute long-standing plans for data-management.[72]

A lapse in self-regulation?

Throughout history, medicine and science have evolved informal systems for censuring risky conduct. Renée Fox and Judith Swazey describe how cardiac surgeons established various moratoria as they encountered difficulties transplanting human hearts. Charles Bosk has described a series of practices and procedures by which surgeons regulate technical and moral performance.[73] Other notable examples are the ways polio-vaccine researchers established and enforced an "informal morality" concerning the initiation and conduct of polio-vaccine studies.[74] And when techniques for splicing DNA were first discovered, top scientists led a moratorium until safety could be established.[75]

Such informal mechanisms for regulating professional conduct seem to have failed gene transfer. Wilson's reputation as a risk taker preceded him (see Chapter 9 for discussion of one early trial that cemented Wilson's reputation), as did reports of adenoviral toxicities. The same month that Gelsinger died, Guenther Cichon and collaborators in Berlin published an article showing hematological toxicities in rabbits after intravenous administration of adenoviral vectors.[76] Cichon later told reporters that administering such vectors to patients like Gelsinger "would never be justified … I am not the only one who thinks this way. We do not understand why [researchers] are taking these risks."[27] Plough-Schering scientists had reportedly observed worrisome adverse events in trials administering adenoviral vectors to the liver of patients with metastatic cancer.[77] According to news reports, several volunteers experienced rapid declines in blood pressure, and one suffered a stroke.**

Wilson reports that no one approached him with concerns about his widely publicized studies,[80] and following Gelsinger's death, several leading researchers expressed regret to that effect. Said David Baltimore, whose discoveries led to the development of gene transfer, "A number of us are asking 'what the hell are we doing putting these things in people?'."[81] Inder Verma, a leading gene-transfer researcher, similarly asked "why didn't we stand up" and raise concerns at meetings?[82] There were many missed opportunities along the way for researchers to warn Wilson about accumulating evidence of adenovirus toxicity.[80] Verma, who holds important gene-transfer patents and serves as an adviser to several biotechnology firms, seemed to answer his own question when he asserted that such warnings would not have been taken seriously because they would have appeared financially motivated.[83]

The challenge of risk classification

As the Gelsinger case illustrates, risks lodge at multiple points in the clinical trial process, and accidents often involve more than hardware failures. The OTC trial provides an opportunity to examine several sets of challenges to the control of risk in translational research. The first is, at its core, a problem of classification. Which are the most important risks to worry about? The physical ones, which are the primary focus of risk deliberations, or the social and institutional ones through which physical risks become manifest? And how should the latter be incorporated into the assessment and management of risk?

Translational trials are governed by a complex and rapidly evolving web of interests – some of which are financial – that complicate risk management. For example, as in the OTC study, lead investigators are frequently named as inventors on patents. The viability of a sponsoring biotechnology company will often depend on the study's timely execution. Professional imperatives further escalate risks. Rewards for investigators who launch first studies are considerable, even where results are negative or unconvincing. The expectations of treatment-refractory patients, and organizations that promote disease research, further prod investigators and institutions toward risky behavior. Add to this mix the overburdened, all-volunteer IRB, straining to decipher a prolix trial brochure while serving its employer.

** Curiously, these go unmentioned in a 2002 study reporting results from this study. A *PubMed* search of SCH 58500 OR ACN 53 shows two published studies involving hepatic administration of the vector. Neither mentions stroke.[78,79]

The tendency in clinical research and human protections has been to conceptualize risk in technical, mono-causal terms. Thus, the US IRB Guidebook identifies two sources of harm that IRBs should focus on: "interventions used in a study, as well as its design features."[84] RAC deliberations typically revolve around questions about vector safety, preclinical data, and study procedures. And informed consent risk descriptions enumerate the possible harms resulting from deliberate, physical perturbations ranging from the trivial (e.g. venipuncture) to the potentially lethal (e.g. off-target effects of vectors). Review, management, and disclosure practices in clinical trials are thus poorly attuned to the types of institutional, systemic, and social factors that doomed the OTC study.

The sociology of risk highlights the limitations of such policies. Brian Wynne, for example, has argued that technical risk assessments embed various assumptions about human and institutional behavior. In 1977 public hearings about a planned nuclear waste reprocessing facility at Sellafield (Windscale, a site in the north-west of England), officials presented seemingly impartial and detailed calculations of the facility's hazards that should have allayed any rational person's fear of the facility. According to Wynne, however, opponents challenged the estimates (vainly, as it turned out) on the grounds that they had not factored in the possibility that the facility might later expand its capacity (and hence, process a larger volume of nuclear waste); nor did they consider the reliability of regulatory oversight mechanisms, which in the previous two decades had been problematic.[85] The technical risk review process used by authorities "silently and systematically deleted ... questions about institutional commitment, behavior, and trustworthiness, as if they had nothing to do with risk."[86]

Though the objective probability of adverse outcomes clearly depends on the validity of any number of social assumptions, technical approaches to risk rarely examine the social assumptions on which they are premised.[86] In the case of the OTC study, reviewers at the RAC assumed that investigators would conscientiously follow their protocol and report adverse events (assumptions about individual behavior). They seem to have trusted that NUCDF's endorsement carried moral and scientific force (an assumption about organizational competence and political legitimacy). Perhaps the RAC assumed, despite concerns about Wilson's research style, that he and the other investigators were capable of managing the competing demands of patient welfare and scientific value (an assumption about judgment), or that the University of Pennsylvania's IRB system would rein in overzealous investigators (an assumption about institutional performance).

These observations invite a series of recommendations for risk review in translational research. The assessment, management, and disclosure of risk would probably function better if the above assumptions were brought to the fore rather than delicately elided. First, review committees should avoid framing their risk assessment restrictively as a technical analysis. To that end, review committees should gather more and higher-quality social information about a study. The current practice within gene transfer is for the OBA to seek answers to fifty or so items for safety assessment. Of these, forty-two items pertain to study and vector design, and only a few directly address the kinds of concerns I have in mind. For example, one item asks about the qualifications of study personnel, a second about the adequacy of study facilities, and a third about how investigators plan to solicit informed consent. One question grazes some of the issues raised in the OTC study: it asks how investigators will ensure open communication should they seek to protect any aspects of the study as intellectual property. Yet in my experience reviewing RAC meetings, answers to these items rarely provoke sustained discussion. Reviewers might go beyond this and seek information about an investigator and their institution's professional and financial interests, their previous performance records, their reputation among colleagues and patients, and whether preclinical safety and efficacy studies have been published in peer review journals.

Disclosure policies would be greatly improved if volunteers had better knowledge about the social process of cutting-edge clinical research. Arguably, what the Gelsingers lacked was not information about adenoviral toxicity, but a more sophisticated appreciation of the diverse and competing interests arrayed around translational research studies. Thus, I have argued elsewhere that IRBs and investigators should be more open-minded about causality when they decide which risks to disclose, and how. For the foreseeable future, all gene-transfer studies should seriously consider disclosing that, in 1999, a volunteer died after participating in a gene-transfer trial. Such a disclosure might be followed up with the statement "Many gene transfer researchers believe the field learned from this tragedy and has fixed its problems, but you might still have concerns. You should discuss this with a third party." In addition, translational trials should supplement the consent process with a frank description of the incentives that motivate clinicians, institutions, and drug companies to pursue research, along with examples of trials that have succeeded and failed to protect the interests of volunteers. Lastly, disclosure statements should emphasize the socially embedded nature of risk. For example, when studies involve a-priori levels of risk that are contentious (and RAC minutes reveal the OTC trial to be one clear example), consent disclosures should avoid opaque statements like "the study agent has never been tested in human beings," and instead offer language that exposes the fallibility of scientific judgments (e.g. "The study agent has never been tested in human beings. We think we know enough about the study agent to test it in people like you. Some scientists disagree, however.").[87]

Normalizing risk

The OTC debacle also provides an occasion to consider how risks are propagated by systems and organizations rather than renegade individuals (this will be further explored in the next chapter). Charles Perrow's *Normal Accidents* argued that technological disasters are an inevitable by-product of increasingly complex systems that create – and manage – risk. In attempting to prevent future accidents, Perrow urged risk managers to look beyond "operator error" to system-wide deficiencies that cause actors or warning systems to fail.[88]

More recently, Diane Vaughan's analysis of decision-making behind the disastrous 1986 space shuttle *Challenger* launch decision contains a number of themes that are echoed in the OTC study. Vaughan describes how various individuals and organizations responsible for detecting and processing danger signals grew increasingly tolerant of risk over time. Small shortfalls in technical performance that might have prompted corrective action came to be seen as a characteristic of routine operation; well-established strategies and norms of risk management came to be viewed as discretionary. One example of such "normalization of deviance" she describes is how management pressures at NASA shifted the burden of proof among NASA engineers from those claiming safety to those warning against shuttle launch.[89] Similarly, my experience with IRBs leads me to believe that the burden of proof is often on doubters rather than those proposing a study.

The OTC episode is rife with instances in which deviance was tolerated during the run-up to Gelsinger's death: the decision to undertake the study in more or less healthy volunteers chafed at what is arguably the most enduring and uncontroversial tenet of medical ethics, "first do no harm." The RAC put aside concerns about biphasic toxicity curves and toxicities observed in earlier studies. Some gene-transfer researchers harbored misgivings about the vector, the protocol, and its lead investigator. The University of Pennsylvania waived its conflict-of-interest policy to enable Wilson's involvement. The study team deliberately strayed from their protocol. Adverse-event reporting systems were far from optimal.

Translational research ethics might therefore benefit from a series of refinements in risk management. First, oversight committees need to develop mechanisms that allow reviewers and peers to come forward with social and contextual safety concerns. The RAC's openness, perhaps one of its greatest virtues, is unfortunately also a hindrance to expressing concerns that implicate the moral character of investigators or the political legitimacy of patient advocate groups. Even where deliberations are closed, individuals might feel uncomfortable raising such issues. Bodies like the RAC or IRBs should provide mechanisms for reviewers to voice social concerns anonymously; in certain situations, closed-door deliberations might be necessary (of course, this introduces new risks, like compromising public accountability).

Second, review committees (and for that matter, research teams) should make liberal use of outside experts who are more likely to bring an independent perspective on various technical issues, and less likely to have their thinking shaped by the culture of a review committee or the organizational imperatives of the research team. Thus, Ph.D. scientists might provide critical feedback where decisions about an early-phase study are clinically motivated. Investigators pursuing alternative strategies might provide appropriate balance in interpreting safety and preclinical studies. In addition, such committees should solicit dissenting viewpoints.

Third, oversight bodies like the RAC, and perhaps professional organizations like the American Society of Gene Therapy, should seek advice from persons who are more likely to be attuned to the social dimensions of risk management. Two disparate classes of professionals come to mind. The first are research nurses, who mediate many transactions between ambitious clinical investigators and research volunteers. The second are social scientists who study organizational and social correlates of risk.

Defining acceptable risk

A third sociological insight – arising again from the work of Charles Perrow – is that catastrophes are inevitable where science is uncertain, technologies complex, and organizations multifaceted. How then, do we decide what level of risk and uncertainty is acceptable? One of the most controversial aspects of the OTC episode was the decision to enroll medically stable volunteers. In basic science-driven research areas like gene transfer, where protocols are often aimed primarily at validating a technological platform rather than curing a particular disease, conditions that present the best opportunity for gathering knowledge offer a less favorable balance of risks and benefits for subjects. How should the appropriate balance between scientific benefit and medical risk be struck?

In the OTC study, the question was formulated in terms of whether to pursue the studies in medically stable, adult volunteers. In other trials, the question takes the form of deciding whether risks are acceptable in study procedures undertaken for strictly scientific reasons.[††]

[††] The standard approach to analyzing risk in clinical trials involves dividing interventions into those with a therapeutic warrant and those performed strictly for scientific purposes.[90] A phase 1 trial, for example, might involve administering a study drug to a research subject (potentially therapeutic) and then collecting organ biopsies to determine the study drug's biodistribution (strictly scientific). Risks of the former are deemed "acceptable" if an intervention could reasonably be considered by clinicians to be consistent with the best medical interests of the subject. Risks of non-therapeutic study procedures are evaluated separately on the basis of whether their scientific value justifies risks imposed on the subject. Component analysis provides a useful tool for systematizing research risks. Nevertheless, it

Such questions have particular significance for translational trials, which aim primarily at gathering basic, scientific information rather than demonstrating clinical efficacy. Debate over the OTC study's subject selection approach thus exemplifies a recurrent question in translational research.

In theory, at least, some commentators appear to suggest that if the value of scientific information were sufficiently great, the level of risk investigators can ask non-vulnerable subjects to endure has no boundaries.[91] Yet many would instinctively reject this utilitarian formulation. Unfortunately, however, research policy has yet to articulate clear standards for defining acceptable risk.[92] One approach that has been explored in other types of medical research[93] might involve numeric standards: policy makers might establish a threshold of risk acceptability, such as one projected serious adverse event per thousand exposed persons. A moment's reflection shows the numeric strategy as conceptually flawed. First, actuarial risk information will not be available for many research activities, particularly in translational research. Second, generic-risk estimates do not reflect the fact that numeric risks of medical procedures frequently depend on the skill and experience of the clinician or on the medical condition of the patient. Third, persons value risk differently; even if precise probabilities and confidence intervals for adverse events are available, magnitudes for harms defy simple quantification. What a committee of policy makers call "acceptable risk" might be considered foolish by a class of situated study volunteers.

An alternative strategy for defining acceptable risk is analogical. Every society allows its members to engage in risky activities that morally resemble research. For example, North Americans often allow (older) children to knock on strangers' doors to raise money for a charity, despite the possibility that our neighbors might wield axes.[94] Or we allow volunteer firefighters to commit hazardous altruistic acts like charging into burning buildings. Why not calibrate research standards to intuitions embodied in existing societal practices?[94,95] The analogical strategy is perhaps more appealing than the numeric one, because it encourages ethicists, investigators, and others to consider precisely what it is we value about different activities rather than obscure these values behind seemingly objective numeric estimates. But it too suffers serious weaknesses. For example, once we decide that research participation is analogous to firefighting, what next? Either researchers design studies that involve dispatching subjects to burning buildings, or they need some instrument for rendering firefighter and research risks comparable. The simplest approach would involve consulting actuarial tables. But this simply recapitulates shortcomings in the numeric approach. Secondly, the analogical strategy seems to displace, rather than resolve, questions of risk. Instead of examining whether a set of hazards is acceptable, ethicists and investigators would need to deliberate over whether research studies were morally similar to firefighting. There is no reason to think that such debates would be any more tractable than those about acceptable risk. A third weakness concerns the moral defensibility of reference points. The fact that North Americans tolerate a certain level of risk for volunteer firefighters says little about whether they *should* tolerate that risk level. As it turns out, when a volunteer firefighter dies in the line of duty, questions inevitably follow that involve the moral uncertainty of risk set-points: was the volunteer exposed to excessive risks because her training and supervision were inadequate? Was the firefighter forced to take risks because of a staffing shortage, and if so, shouldn't the city raise taxes to

---

provides no guidance on how risk and benefits in the non-therapeutic arm should be balanced; as conceded by Weijer, it would seem to suggest that if the value of scientific information were sufficiently great, the level of risk investigators can ask non-vulnerable subjects to endure has no boundaries.[91]

pay a professional force? Decisions about risk acceptability often reflect an imperfect political process rather than careful moral deliberation.

A third approach – the one I believe has particular utility for translational research – is procedural. Rather than advocating substantive guidelines about acceptable risk, this approach would accept the outcome of any risk assessment provided its process incorporates the views of all relevant interests to the research. Often, parties to the risk-assessment process will have diverging views about a study's risks. Some investigators, enthused with their research, might have a risk tolerance that many prospective subjects would regard as abusive or degrading for persons with their disease. Some IRB members, reacting to the investigator's zeal, might advocate restrictive risk levels. Prospective subjects enthused with a particular line of research could rightly see the ethicist's caution as an infringement of their autonomy, and Robert Levine recommends erring on the side of autonomy when such conflicts occur.[96] But these same subjects might also encourage researchers to perform manipulations that physician–investigators feel would abrogate their duty to "do no harm." Clearly, investigators have no positive duty to commit research acts on willing subjects. Ideally, decisions about subject selection models and risk–benefit ratios should involve deliberations between investigators, ethics committees, and prospective research subjects.

Because IRBs evaluate protocols submitted by investigators, views of the former two categories are well represented in risk deliberations. With rare exception, however, the views of prospective subjects are only indirectly represented by IRB community representatives and ethicists.[‡‡] Empirical as well as testimonial evidence reveals both as flawed surrogates.[97] To develop procedurally defensible risk decisions, translational-research processes will need to devise systems for incorporating the perspectives of those who will be invited into research protocols.

The most obvious means of doing so would be to invite the participation of patient advocates in the planning and review of study protocols. As will be discussed in Chapter 5, many disorders investigated in translational trials involve well-organized and scientifically sophisticated advocacy organizations. Such groups offer two opportunities for improving the procedural rigor of risk review. First, they can participate in designing studies by establishing criteria for initiating trials,[98] helping to decide whether a trial's risks and benefits are appropriately balanced, and expressing concerns about a research protocol publicly. Second, research advocates can contribute to ethical review of proposed protocols. Where translational trials involve contentious levels of risk, IRBs or the RAC should invite disease advocates on an ad hoc basis to participate in review.[99]

Group participation in research design and risk review has several important shortcomings. Perhaps the most obvious is the question of representation: which organization is best suited to vouch for prospective subjects? Persons with hemophilia, for example, might be represented by the World Federation of Hemophilia, the National Hemophilia Foundation (NHF), or the Committee of Ten Thousand.[§§] All have very different agendas and orientations: the first emphasizes the unmet needs of persons in the developing world, the second is primarily involved in research advocacy, and the third represents persons infected with

[‡‡] RAC's membership has also historically included at least one person to represent the views of prospective subjects. One example of a member representing this viewpoint is Abbey Meyers, president of the National Organization of Rare Disorders.

[§§] This group was founded in the late 1980s in the wake of HIV contamination of the blood supply. The group practices a more overtly politicized brand of disease advocacy.

HIV from blood products. Though many persons with hemophilia express strong support for gene-transfer research, many others harbor deep distrust of groups like the NHF, which many believe did not respond effectively as HIV spread to the blood supply in the 1980s.[100,101] Similar divisions characterize other disease constituencies as well.[30,102] A second problem is that advocacy organizations often serve competing interests.[103] Disease advocacies derive a significant fraction of their budget from pharmaceutical and biotechnology companies. Many persons with hemophilia, for example, believe that organization ties to the blood-product industry prevented an aggressive response to the threat of HIV. Another perennial conflict within advocacy organizations concerns the competing constituencies of present patients (whose needs center around services and support) and future patients (whose needs are generally seen to be served by medical research). Advocacy organizations are generally not run on a democratic model, and the balance they strike between these various interests is often more a result of fund-raising opportunities and political calculation. Related to this is a third major limitation: groups are often more tolerant of risks than individual members of their constituency.[104] This might be a particular concern where advocacy organizations are swept up by fast-breaking scientific developments. Perhaps this partly explains how the NUCDF came to endorse the OTC study.

In anticipation of a more procedural orientation to risk review, two recommendations can be offered to address the problems discussed above.[§§] First, investigators and ethics committees should be energetic about understanding patient communities. Though they should solicit the participation of advocacy organizations in study planning and review, they should also be prepared to interrogate their views with the aim of appraising their legitimacy as surrogates for prospective subjects. Second, in addition to seeking the input of major advocacy organizations, which often have a more "future-patient" orientation, investigators and ethicists should also solicit opinion from the more "present-patient" centered support groups.

Skeptics might object that including patient advocates in study design and risk assessment might further encumber the research process. Two responses can be offered. First, a rebuttal: investigators who claim that their protocol advances the medical interests of a class of patients have an obligation to understand what those interests are. Second, I concede that my recommendations pertain to protocols that involve levels or types of risk that might be controversial. Protocols that represent minor iterations on previously approved protocols need not solicit the views of patient advocates (unless there are grounds for thinking that earlier protocols are themselves controversial). As well, patient advocates might reasonably decline to contribute to the risk-review process. I would argue that there are four circumstances where the views of patient advocates should be sought routinely. First, whenever scientific considerations lead investigators to design protocols enrolling medically stable patients in trials of highly novel agents (e.g. most hemophilia gene-transfer studies). Second, wherever scientific considerations necessitate the use of unusually risky or burdensome study procedures (e.g. a study involving a large number of invasive biopsies). Third, whenever a protocol involves a study intervention whose properties are poorly understood (e.g. the first application of a novel vector in human subjects). Fourth, whenever protocols raise morally complex risk–benefit tradeoffs (e.g. survival vs. major risk of malignancy) or ethically contentious risks (e.g. vertical transmission).

§§ Recommendations for how disease advocates might conduct themselves will not be discussed here; interested readers should see Dresser R. *When Science Offers Salvation*.[99]

Table 3.1 Recommended practices for translational trials involving contentious levels of risk

(1)	Revisit / entertain key social assumptions that underpin risk assessments
(2)	Create mechanisms for reviewers or others to express safety concerns anonymously
(3)	Solicit perspective from scientists/physicians outside the immediate research area
(4)	Solicit dissenting viewpoints from scientists/physicians
(5)	Seek advice from persons knowledgeable about social practices in translational trials
(6)	Include diverse disease-advocate viewpoints in protocol development and review
(7)	Use disclosure approaches during consent that emphasize the fallibility of social practices

Conclusions: risk and translational research

In what way do translational studies present distinctive challenges to the assessment and management of risk in human research? My account of the University of Pennsylvania OTCD trial offers several answers. First, translational research trials bring together a potent combination of divergent interests, including ambitious and professionally invested researchers, disease advocacy organizations, biotechnology companies, and elite research institutions that might be tempted to bend rules to retain star status.[105] Add to this mix zealous news media and study volunteers who are either desperate, or strongly identified with their illness such that they are willing to endure substantial risk to advance research. At numerous points these interests amplify, propagate, or introduce risks in ways that might not be anticipated by ethical and regulatory oversight systems that conceptualize risks as physical, probabilistic phenomena.

Second, translational research is characterized by shifting moral, regulatory, and scientific landscapes. In the period between the protocol's approval and Gelsinger's demise, financial conflict-of-interest policies at academic institutions had undergone significant changes, the RAC's approval authority was eliminated, relationships between the FDA and the RAC became strained, and new generations of adenoviral vectors became available. Wilson had only recently moved to the University of Pennsylvania, which had established one of the first centers of its kind, the Institute for Human Gene Therapy. What matters in terms of safety, law, policy, and ethics is not yet settled in new research areas. This absence of clear guidance and lines of accountability leaves new fields susceptible to major mishaps.

A third major factor is the profound technical uncertainty surrounding novel interventions. Under such conditions, scientific and clinical claims are much less constrained by evidence, such that neither claims of safety nor claims of hazard have much support. When such conditions of under-determination prevail, technical decisions are largely driven by worldviews, values, affiliations, and desires.

This chapter presented risk as a complex, emergent, and collective phenomenon rather than as the consequence of physical acts performed by generic actors. Recommendations stemming from this analysis are summarized in Table 3.1. In the chapter that follows, I will extend this analysis to an episode where deficiencies in conduct and practice like those catalogued above are much less apparent: the second ever gene-transfer-induced death.

References

1. Douglas M. *Risk and Blame: Essays in Cultural Theory*. New York, NY, Routledge, 1992.
2. Assessment of adenoviral vector safety, and toxicity: report of the National Institutes of Health Recombinant DNA Advisory Committee. *Hum Gene Ther* 2002; **13** (1): 3–13.
3. Nasto B. Questions about systemic adenovirus delivery. *Mol Ther* 2002; **5**(6): 652–3.

4. Miller HI. Health care: Regulatory overdose. *Hoover Digest: Research and Opinion on Public Policy* 2000; 3.
5. Smaglik P. Investigators ponder what went wrong. *The Scientist* 1999; **13**(21): 1.
6. Savio W, Verma I. ASGT Opinion. Editorial. *ASGT Press Release*, 2000.
7. Mitchell S. Gene therapy benefit said to outweigh risk. *United Press International* 2002 October 10.
8. Crystal RG, McElvaney NG, Rosenfeld MA, Chu CS, Mastrangeli A, Hay JG, *et al.* Administration of an adenovirus containing the human CFTR cDNA to the respiratory tract of individuals with cystic fibrosis. *Nat Genet* 1994; **8**(1): 42–51.
9. Recombinant DNA Advisory Committee (RAC). Minutes of Meeting, Dec 4–5 1995.
10. Macilwain C. NIH, FDA seek lessons from hepatitis B drug trial deaths. *Nature* 1993; **364**(6435): 275.
11. Marwick C. NIH panel report of 'no flaws' in FIAU trial at variance with FDA report, new probe planned. *JAMA* 1994; **272**(1): 9–11.
12. Recombinant DNA Advisory Committee (RAC). Minutes of Meeting, September 25–26, 2000. Available at: www4.od.nih.gov/oba/rac/minutes/Sept00RAC.
13. Recombinant DNA Advisory Committee (RAC). Minutes of Meeting, September 6–7, 2001. Available at: www4.od.nih.gov/oba/rac/minutes/RAC%20Min%20090601.pdf (last accessed on 7 October 2007).
14. Recombinant DNA Advisory Committee (RAC). Minutes of Meeting, December 8 1999.
15. Masiello SA. Notice of Initiation of Disqualification Proceedings and Opportunity to Explain to Mark L. Batshaw M.D. US Food and Drug Administration. www.fda.gov/foi/nidpoe/n14l.pdf 2000 November 30.
16. RAC, Minutes of Symposium and Meeting, December 8–10, 1999.
17. Nunes F, Furth E, Wilson J, Raper S. Gene transfer into the liver of nonhuman primates with E1-deleted recombinant adenoviral vectors: safety of readministration. *Hum Gene Ther* 1999; **10**(15): 2515–26.
18. BRMAC meeting April 5, 2001, pp. 201–300. Testimony of Joseph P. Salewski.
19. Holmes EW. Of rice and men: Bill Kelley's next generation. *J Clin Invest* 2005; **115**(10): 2948–52.
20. Anonymous. World renowned researcher James Wilson, MD, PhD, to lead University of Pennsylvania Medical Center's gene therapy initiative. *PR Newswire* 1992 October 5.
21. Wilson JM, Engelhardt JF, Grossman M, Simon RH, Yang Y. Gene therapy of cystic fibrosis lung disease using E1 deleted adenoviruses: a phase I trial. *Hum Gene Ther* 1994; **5**(4): 501–19.
22. Raper SE, Grossman M, Rader DJ, Thoene JG, Clark BJ, 3rd, Kolansky DM, *et al.* Safety and feasibility of liver-directed ex vivo gene therapy for homozygous familial hypercholesterolemia. *Ann Surg* 1996; **223**(2): 116–26.
23. Eck SL, Alavi JB, Alavi A, Davis A, Hackney D, Judy K, *et al.* Treatment of advanced CNS malignancies with the recombinant adenovirus H5.010RSVTK: a phase I trial. *Hum Gene Ther* 1996; **7**(12): 1465–82.
24. Flowers CR, Melmon KL. Chapter 6. In: Landau R, Achilladelis B, Scriabine A, eds. *Pharmaceutical Innovation: Revolutionizing Human Health.* Philadelphia, Chemical Heritage Press, 1999; 331–72.
25. Gutterman JU. Clinical investigators: the driving force behind drug discovery. *Nat Biotechnol* 1997; **15**(7): 598–9.
26. Goldstein JL, Brown MS. The clinical investigator: bewitched, bothered, and bewildered – but still beloved. *J Clin Invest* 1997; **99**(12): 2803–12.
27. Nelson D, Weiss R. Hasty decisions in the race to a cure. *The Washington Post Sunday* 1999 November 21.
28. Anderson C. Gene therapy. A speeding ticket for NIH's controversial cancer star. *Science* 1993; **259**(5100): 1391–2.
29. Marshall E. Panel approves gene trial for "normals". *Science* 1997; **275**(5306): 1561.
30. Stockdale A. Waiting for the cure: Mapping the social relations of human gene therapy research. *Sociology of Health and Illness* 1999; **21**(5): 579–96.
31. Garber K. RAC nixes plan to treat retinoblastoma. *Science* 1999; **284**(5423): 2066.
32. Vaughan D. *The Challenger Launch Decision: Risky Technology, Culture, and*

Deviance at NASA. Chicago, University of Chicago Press, 1996.

33. Batshaw ML. *Children with Disabilities*, 4th edn. Baltimore, Paul H. Brookes Publishing, 1997.
34. Gelsinger P. Jesse's Intent. *Citizens for Responsible Care and Research (CIRCARE).*
35. Chen J. Profile: Failure of research scientists to inform patients of risks in experimental gene therapy. *CBS News: Morning News* 2000 February 3.
36. Wilson JM. Method of Treating Liver Disorders (Patent). University of Pennsylvania,. WO97*30167*, United States, 1997.
37. Stolberg SG. Biomedicine is receiving new scrutiny as scientists become entrepreneurs. *The New York Times* 2000 February 20.
38. Gelsinger vs. Trustees of the University of Pennsylvania. In: Sherman, Silverstein, Kohl, Rose and Podolsky, Law Offices, Philadelphia, 2000.
39. Baram M. Making clinical trials safer for human subjects. *Am J Law Med* 2001; **27**(2–3): 253–82.
40. Hensley S. Targeted genetics' Genovo deal leads to windfall for researcher. *The Wall Street Journal* 2000 August 10.
41. Stolberg S. The biotech death of Jesse Gelsinger. *The New York Times* 1999 November 28.
42. Chartrand S. Patents: A gene therapy technique that is seen as easier and less expensive than the main alternative. *The New York Times* 1997 October 6.
43. Kelley WN, Palella TD, Levine M. Viral-mediated Gene Transfer System (Patent). In: The Regents of the University of Michigan, 1997.
44. American Society of Gene Therapy. *Policy of the American Society of Gene Therapy: Financial Conflict of Interest in Clinical Research*. Adopted April 5, 2000. Available at: www.asgt.org/position_statements/conflict_of_interest.php.
45. Department of Health and Human Services. *Financial relationships in clinical research: Issues for institutions, clinical investigators, and IRBs to consider when dealing with issues of financial interest and human subject protection – Draft Interim Guidance.* Department of Health and Human Services, Office of Human Research Protections, 2001.
46. AAMC Task Force on Financial Conflicts of Interest in Clinical Research. Protecting subjects, preserving trust, promoting progress I: policy and guidelines for the oversight of individual financial interests in human subjects research. *Acad Med* 2003; **78**(2): 225–36.
47. AAMC Task Force on Financial Conflicts of Interest in Clinical Research. Protecting subjects, preserving trust, promoting progress II: principles and recommendations for oversight of an institution's financial interests in human subjects research. *Acad Med* 2003; **78**(2): 237–45.
48. Gelsinger P. Committee on Health, Education, Labor, and Pensions. Subcommittee on Public Health, 2000 February 2.
49. Munson R. *Outcome Uncertain: Cases and Contexts in Bioethics*. Toronto, Wadsworth Publishing, 2002.
50. From posting on blog: blog.bioethics.net. Last accessed March 28, 2008. On file with author. http://blog.bioethics.net/2008/01/a-comment-from-paul-gelsinger-on-gene-therapy-and/
51. Kim SY, Millard RW, Nisbet P, Cox C, Caine ED. Potential research participants' views regarding researcher and institutional financial conflicts of interest. *J Med Ethics* 2004; **30**(1): 73–9.
52. Grady C, Horstmann E, Sussman JS, Hull SC. The limits of disclosure: what research subjects want to know about investigator financial interests. *J Law Med Ethics* 2006; **34**(3): 592–9, 481.
53. Gray SW, Hlubocky FJ, Ratain MJ, Daugherty CK. Attitudes toward research participation and investigator conflicts of interest among advanced cancer patients participating in early phase clinical trials. *J Clin Oncol* 2007; **25**(23): 3488–94.
54. Hampson LA, Agrawal M, Joffe S, Gross CP, Verter J, Emanuel EJ. Patients' views on financial conflicts of interest in cancer research trials. *N Engl J Med* 2006; **355**(22): 2330–7.
55. Kaiser J. Clinical trials. Gene transfer an unlikely contributor to patient's death. *Science* 2007; **318**(5856): 1535.
56. Weiss R. Death points to risks in research. *The Washington Post* 2007 August 6.

57. Ethics: Building your strategic tool kit for safe patient accrual. *American Society for Gene Transfer: 11th Annual Meeting*, 2008. Boston, MA, May 28 to June 1, 2008.
58. Kahn J. Informed consent in human gene transfer clinical trials. *Hum Gene Ther* 2008; **19**(1): 7–8.
59. Silber TJ. Human gene therapy, consent, and the realities of clinical research: is it time for a research subject advocate? *Hum Gene Ther* 2008; **19**(1): 11–14.
60. Caplan AL. If it's broken, shouldn't it be fixed? Informed consent and initial clinical trials of gene therapy. *Hum Gene Ther* 2008; **19**(1): 5–6.
61. US General Accounting Office. *Scientific Research: Continued Vigilance Critical to Protecting Human Subjects*. In: Health, Education, and Human Services Division, US General Accounting Office, 1996.
62. US Department of Health and Human Services. *Institutional Review Boards: A Time for Reform*. Department of Health and Human Services, US Office of Inspector General, 1998.
63. Oinonen MJ, Crowley WF, Jr., Moskowitz J, Vlasses PH. How do academic health centers value and encourage clinical research? *Acad Med* 2001; **76**(7): 700–6.
64. Report of independent panel reviewing the University of Pennsylvania's Institute for Human Gene Therapy. *Penn Almanac* 2000; **46**(34), May 30.
65. Steinbrook R. Improving protection for research subjects. *N Engl J Med* 2002; **346**(18): 1425–30.
66. Marshall E. Shutdown of research at Duke sends a message. *Science* 1999; **284**(5418): 1246.
67. Holden C. University of Illinois. Chancellor quits after research shutdown. *Science* 1999; **285**(5436): 2047.
68. Shalala D. Protecting research subjects – what must be done. *N Engl J Med* 2000; **343**(11): 808–10.
69. US Food and Drug Administration. *Points to Consider in the Production and Testing of New Drugs and Biologicals Produced by Recombinant DNA Technology*. US Department of Health and Human Services, 1985 April 10.
70. Pollner F. Gene therapy trial and errors raise scientific, ethical, and oversight questions. *The NIH Catalyst* 2000; **8**(1).
71. Politics & Policy – Gene therapy: NIH admits failure to monitor programs. *American Health Line* 2000 February 2.
72. Walters L. Safety of gene therapy. *Congressional Testimony by Federal Document Clearing House*. 2000 February 2.
73. Bosk CL. *Forgive and Remember: Managing Medical Failure*. Chicago, University of Chicago Press, 1979.
74. Halpern SA. *Lesser Harms: The Morality of Risk in Medical Research*. Chicago, University of Chicago Press, 2004.
75. Wright S. *Molecular Politics: Developing American and British Regulatory Policy for Genetic Engineering 1972–1982*. Chicago, University of Chicago Press, 1994.
76. Cichon G, Schmidt HH, Benhidjeb T, Loser P, Ziemer S, Haas R, *et al*. Intravenous administration of recombinant adenoviruses causes thrombocytopenia, anemia and erythroblastosis in rabbits. *J Gene Med* 1999; **1**(5): 360–71.
77. Stolberg S. New data released on side effects in gene therapy experiments. *The New York Times* 1999 Nov 20.
78. Warren RS, Kirn DH. Liver-directed viral therapy for cancer p53-targeted adenoviruses and beyond. *Surg Oncol Clin N Am* 2002; **11**(3): 571–88, vi.
79. Atencio IA, Grace M, Bordens R, Fritz M, Horowitz JA, Hutchins B, *et al*. Biological activities of a recombinant adenovirus p53 (SCH 58500) administered by hepatic arterial infusion in a Phase 1 colorectal cancer trial. *Cancer Gene Ther* 2006; **13**(2): 169–81.
80. Nelson D, Weiss R. Gene therapy study proceeded despite safety, ethics concerns. *The Washington Post* 1999 November 21.
81. Monmaney T. Gene therapy called too risky; Science: Caltech's Baltimore, a leader in genetics research, says technique should not be tested on people. *Los Angeles Times* 2000 Feb 19.
82. Barinaga M. Asilomar revisited: lessons for today? *Science* 2000; **287**(5458): 1584–5.
83. Friend T. Conflicting interests compromise safety. *USA Today* 2000 February 22.
84. Penslar RL. Insitutional Review Board Guidebook. US Department of Health and Human Services, Office for Human Research Protections, 1993.

85. Wynne B. *Rationality and Ritual: The Windscale Inquiry and Nuclear Decisions in Britain*. Chalfont St Giles, Bucks, The British Society for the History of Science, 1982.
86. Wynne B. Risk and Social Learning: Reification to Engagement. In: Krimsky S, Golding D, eds. *Social Theories of Risk*. New York, Praeger Publisher, 1992; 275–300.
87. Kimmelman J. Acknowledging fallibility: Risk disclosure and informed consent. *Bioethics* (In press).
88. Perrow C. *Normal Accidents: Living with High-Risk Technologies*. Princeton, NJ, Princeton University Press, 1999.
89. Vaughan D. *The Challenger Launch Decision: Risky Technology, Culture, and Deviance at NASA*. Chicago, University of Chicago Press, 1996; 338–9.
90. Weijer C, Miller PB. When are research risks reasonable in relation to anticipated benefits? *Nat Med* 2004; **10**(6): 570–3.
91. Weijer C. Thinking clearly about research risk: implications of the work of Benjamin Freedman. *IRB* 1999; **21**(6): 1–5.
92. Emanuel EJ, Wendler D, Grady C. What makes clinical research ethical? *JAMA* 2000; **283**(20): 2701–11.
93. Wendler D, Belsky L, Thompson KM, Emanuel EJ. Quantifying the federal minimal risk standard: implications for pediatric research without a prospect of direct benefit. *JAMA* 2005; **294**(7): 826–32.
94. Wendler D. Protecting subjects who cannot give consent: toward a better standard for "minimal" risks. *Hastings Cent Rep* 2005; **35**(5): 37–43.
95. London AJ. Reasonable risks in clinical research: a critique and a proposal for the Integrative Approach. *Stat Med* 2006; **25**(17): 2869–85.
96. Levine RJ. *Ethics and Regulation of Clinical Research*, 2nd edn. New Haven, Yale University Press, 1988.
97. Dresser R. *When Science Offers Salvation: Patient Advocacy and Research Ethics*. Chapter 6. New York, NY, Oxford University Press, 2002.
98. National Hemophilia Foundation. *Recommendations for conducting gene transfer clinical trials in persons with bleeding disorders*, revised November 2002. MASAC *Document #137*. www.hemophilia.org/NHFWeb/Resource/StaticPages/menu0/menu5/menu57/masac137.pdf; 2002.
99. Dresser R. *When Science Offers Salvation: Patient Advocacy and Research Ethics*. New York, NY, Oxford University Press, 2002.
100. Keshavjee S, Weiser S, Kleinman A. Medicine betrayed: hemophilia patients and HIV in the US. *Soc Sci Med* 2001; **53**(8): 1081–94.
101. Bayer R. Blood and AIDS in America: The Making of a Tragedy. In: Feldman E, Bayer R, eds. *Blood Feuds: AIDS, Blood, and the Politics of Medical Disaster*. New York, Oxford University Press, 1998; 19–45.
102. Epstein S. Impure science: AIDS, activism, and the politics of knowledge. Berkeley: University of California Press, 1996.
103. Berenson A. In drug-aid foundations, a web of corporate interests. *The New York Times* 2006 April 8.
104. Melton GB, Levine RJ, Koocher GP, Rosenthal R, Thompson WC. Community consultation in socially sensitive research. Lessons from clinical trials of treatments for AIDS. *Am Psychol* 1988; **43**(7): 573–81.
105. Lo B, Wolf LE, Berkeley A. Conflict-of-interest policies for investigators in clinical trials. *N Engl J Med* 2000; **343**(22): 1616–20.

Chapter 4 Taming uncertainty: risk and gene-transfer clinical research

One clear bag of fluid

In September 2001, one-year-old Rhys Evans became the poster child for gene transfer. The Welsh boy had been diagnosed with X-linked Severe Combined Immune Deficiency (X-SCID) after months of infections, oxygen tents, intensive care, and finally, an isolation room. His parents – one a teacher, the other a pipe fitter – were offered two options by his caregivers at the Great Ormond Street Hospital in London: bone-marrow transplantation plus a brutal regime of chemotherapy (to allow the new tissue to "take"), or an experimental procedure – gene transfer – that had shown success in two infants at Paris's Necker Hospital. Fearing the potentially lethal consequences of chemotherapy, the parents elected for gene transfer, and in July Evans received what his mother recorded in her diary as "one clear bag of fluid."[1] Within weeks, immunological parameters began normalizing, and as Rhys approached his second birthday he was discharged from the hospital. Today, Evans leads a normal life – aside from occasional public appearances for charity, and lots of medical monitoring.

The latter has proven especially important in Evans's case because, across the English Channel, several of his X-SCID peers who underwent the same procedure were not as fortunate. Under the leadership of immunologists Alain Fischer and Marina Cavazzana-Calvo, several children had been cured in a protocol initiated in 1998. But by the early summer of 2002, the French team detected a never-before observed lymphoproliferative disorder in one of their patients. By December, a second patient began showing identical indications. Two years later, a third patient developed the disorder. In 2007, a fourth case was observed, and then a fifth (this time in the UK group) was observed in 2008. In three instances, the abnormal cells showed a similar molecular signature: the retroviral vector had integrated near a proto-oncogene, LMO2, associated with T-cell lymphoblastic leukemias.[2]

Insertional mutagenesis – that is, genetic injury caused by the insertion of a genetic element – had long been predicted theoretically. After all, the vector used in the protocol is derived from the "Moloney *leukemia* virus," [my italics] and in 1995, the Orkin–Motulsky report warned that "because clinical experience is still so limited, it is not possible to exclude long-term [sic] adverse effects of gene transfer therapy, such as might arise from mutations when viral sequences randomly integrate at critical sites in the genome of somatic cells. It must be noted that multiple integration events resulting from repeated administration of large doses of retroviruses theoretically pose a risk for leukemic transformation."[3]

The role of insertional mutagenesis in triggering the lymphoproliferative disorders remains unclear.* Nevertheless, until 2002, insertional mutagenesis was considered a purely

---

* The team that led the X-SCID trial has argued that insertional mutagenesis is the primary cause of the leukemia cases. However, other research teams have suggested that transgene overexpression may have played a role.[4–6]

speculative concern. A typical informed consent document warned: "one important theoretical risk is the possible development of cancer..." and then qualified it with "the development of cancer has not been observed in any of the animals or patients treated with this type of disabled virus. It is our opinion that the risk of developing a cancer as a result of this procedure is extremely low."[†] Leukemias that occurred in animals involved retroviruses that were capable of propagation, and could thus disrupt multiple sites within a genome. In contrast, gene-transfer protocols used small amounts of a nonviable vector. Statistically, the chances of activating an oncogene seemed remote, at least if one assumed that vectors integrated at random locations within the human genome.

The nature of the problem

The X-SCID leukemias were one of a series of adverse events that revealed unpredictable properties of gene-transfer agents. The previous chapter described encounters with adenoviral toxicity in 1994[7] and 1999.[8] Subclinical problems arose again using an even more disabled adenovirus in a 2001 hemophilia trial, which was stopped after a transient rise in serum transaminases and thrombocytopenia in the trial's first subject.[9] Unexpected adverse events occurred with other vector platforms as well, though without major or irreversible harm. One example was the surprising immunogenicity of AAV vectors in another trial involving hemophilia. The same trial also detected vector in the semen of volunteers, raising concerns about horizontal and vertical vector transmission. The former had not been predicted in preclinical testing; the latter was only variably predicted from animal studies.[10]

Is gene transfer intrinsically more dangerous than typical small-molecule drugs? Does participation in a gene-transfer clinical trial involve greater risk? The convention within research ethics is to define risk as the product of an unfavorable outcome's probability and magnitude. Using this definition, gene transfer would not appear to be any riskier than treatments typically offered to persons with a given illness. For example, a 2005 study examining the risk of participating in a phase 1 cancer study found that, of 112 patients participating in gene-transfer studies, none died from toxic events. The figure for cytotoxic chemotherapy – the most commonly tested class of cancer drugs – is around 0.66 percent.[‡] For children with X-SCID who cannot receive optimal treatment (that is, a matched bone marrow donation), best-case median three-year survival is 64 percent.[12] Compare that figure to a crude estimate of 85 percent success in 20 attempts using recent gene transfer in children with X-SCID.[13]

Rather, what distinguishes the risks of gene transfer from those for conventional drugs is their level of complexity and uncertainty. Systems for anticipating, detecting and managing gene-transfer risks are inchoate. Adverse events like those described above remain poorly understood. Despite scores of trials involving retroviral vectors, there is not yet any widely accepted system for quantifying the risks of insertional mutagenesis.[14] Years after the above X-SCID studies, an explanation for the dramatic safety differences in the London and Paris studies continues to elude researchers.

[†] The identity of the specific consent document is suppressed to protect the privacy of the investigator team. The protocol from which this derived was submitted to RAC between 1995 and 2000.

[‡] It should be noted that the number of gene transfer subjects assessed for toxicity in this study would not necessarily register a death rate of 0.66 percent. However, we can at least make an educated guess from these data that the risks of gene transfer are not grossly greater than for cytotoxics which, by design, are poisons.[11]

This chapter concerns the ethical response to risk and, more specifically, uncertainty. I begin by describing some of the reasons why gene-transfer protocols involve high degrees of risk uncertainty. I will then examine the moral response to uncertainty in terms of how risks are evaluated, minimized, and managed inside and outside of trial protocols.

Uncertainty and gene-transfer interventions

Uncertainty about risk in gene-transfer experiments is partly a product of several technical factors. These are enumerated below.

Passive vs. active compositions

Conventional agents like small-molecule drugs are passive compositions of matter, and are thus entirely dependent on the host for distribution and metabolism.[§] Gene transfer and other novel interventions often involve active agents that execute a biological program. Minimally, vectors install their genome within a recipient's cells and drive expression of a transgene. But other protocols involve agents that do quite a bit more. Some involve viruses that replicate conditionally (that is, they propagate only in selected cell types), and others involve "living" interventions (one example involves the use of bacteria to deliver an anti-inflammatory cytokine, IL-10, to the gut of persons with inflammatory bowel disease).[16] Even where viral vectors have only a minimal capacity to carry out a biological program, three additional concerns remain. The first is the possibility that fully functional viruses might contaminate vector preparations during the production process, thereby exposing volunteers to the risk of infection. Second is the possibility that vectors might "recombine" with related viruses infecting the host, and recover their capacity to replicate. The third concern is that the vector might trigger a reactivation of viruses that are otherwise dormant in the human genome.[17]

Mode of action

Gene-transfer agents function by a different mechanistic pathway than typical drugs. First, whereas the latter affect gene expression, gene-transfer agents directly intervene in the genetic circuitry. However, the pathways by which genes act are not always well understood, and genetic alterations occasionally lead to unexpected outcomes. For example, mouse embryos genetically engineered to express higher levels of the growth hormone protein develop into larger but otherwise normal mice. The same genetic alteration in pigs, however, weakens the animals' resistance to stress and causes movement difficulties.[18] Gene-transfer techniques might produce gene products in tissues that are not normally exposed to a particular protein.[¶] Or improper regulation of a transgene expression can cause deleterious effects.[20] In addition, some gene-transfer interventions target novel mechanisms or biological pathways. Again, viral oncolytic therapy provides one example. Another is a recent Parkinson's disease trial that attempted to convert a structure in the brain that stimulates dopaminergic neurons (the subthalamic nucleus) into one that could inhibit their action.[21]

---

§ Of course, the use of active agents is not entirely unprecedented. Some vaccines use living, attenuated viruses. One interesting example is the Soviet attempts to use phage to treat bacterial infection.[15]

¶ An example: This can lead to unanticipated consequences, as when researchers observed the emergence of athlerosclerosis in rabbits after using gene transfer to deliver a widely used enzyme for the treatment of thrombolytic disorders, urokinase-type plasminogen activator, to the arteries of rabbits.[19]

Complex composition

Gene-transfer agents are complex in that they simultaneously embody delivery devices (viruses or liposomal particles) and pharmacologic agents (transgenes). Some protocols also involve other procedures that entail risk, such as the use of stem cells, non-myeloablative condition regimes, or transient immunosuppression. Though each might present safety concerns, toxicities can also emerge as a result of cooperative interactions between components. For example, some scientists speculate that the lymphoproliferative disorders observed in the Paris-Necker X-SCID study resulted from the combined effects of the transgene and the vector's insertion near the LMO2 locus.[22] Another interactive risk is the possibility of viral vectors stimulating immune responses against therapeutic transgenes; for some diseases, this might eliminate a standard therapeutic option of using enzyme replacement therapy.[23, 24]

Long-term impacts

Gene transfer often involves stable or permanent genetic alterations. As a result, recipients of gene transfer are potentially exposed to long-term and continuous levels of a transgene product. This has three consequences with respect to risk and uncertainty. First, preclinical data rarely include long-term outcomes and are often not predictive for latent effects. Second, extended exposures increase the likelihood that subtle safety issues might emerge over time. Third, the possibility that toxicities might be irreversible because of an intervention's durability increases the magnitude of negative outcomes.

Species specificity

Biological techniques like gene transfer often show a high degree of species-specific toxicity. For example, mice, the main organism used in preclinical toxicity testing, do not develop infections when exposed to adenoviruses. As a result, they have limited utility for modeling the safety and biodistribution of adenoviral vectors. Though other species can be substituted (e.g. cotton rats or primates), these animals have a different major disadvantage: they rarely develop the types of pathologies being studied in clinical trials.[25]** Like vectors, transgenes can also have species-specific properties if they are directed toward targets unique to human beings. This leads to a paradox for molecular medicine: agents that have greater specificity for a particular target are presumed to be safer, because off-target effects are greatly diminished. However, such agents are more difficult to model in preclinical studies.††

Immune-based toxicity

Many toxicities in gene transfer are immunologically based. The previous chapter describes several instances where potent immunological reactions were triggered by viral vectors. Another example – unproven as of this writing – is the possible role of adenoviral vectors somehow exarcerbating HIV infection in volunteers with higher adenovirus antibody

---

** As stated in a workshop involving leaders in gene transfer research: "Human adenoviruses do not replicate well in other species and most animal populations cannot mimic the human population's antigenic history. Ploegh pointed out that many studies demonstrate that previous exposure to one virus can dramatically affect the response to subsequent viral exposures, even to unrelated viruses. A concerted effort is required to develop better animal models."[26]

†† A report on first-in-human trials, issued in the wake of the TGN 1412 trial, put it as follows: "species-specificity of an agent does not imply that there is always an increased risk in first-in-man trials, but it

titers.[‡‡] This presents at least two difficulties. First, human beings vary widely in how they mount immune responses, and researchers are, at present, unable to predict an individual's immune response. Moreover, the status of the immune system at any given time will depend on a person's prior exposure to related antigens, or whether their immune system has been activated because of a concurrent infection.[29] Second (and related to the species specificity issue described in the previous paragraph), immunotoxicity is highly species-specific, and is thus notoriously difficult to model.[30–33]

Nonlinear dose–response relationships

Gene-transfer agents show a nonlinear dose–response curve for certain vectors.[34] This was discussed in the previous chapter. Briefly, some vectors do not exhibit significantly greater toxicity until the dose reaches a certain threshold, after which toxicity increases precipitously. This poses a significant challenge to predicting safety; when threshold toxicity triggered the death of Jesse Gelsinger, investigators concluded that this, along with "subject-to-subject variation," made it "difficult to evaluate vector safety in a traditional Phase I trial design . . . "[8]

Uncertainty and risk estimation in clinical trials

To be clear: the argument is not that gene transfer is intrinsically riskier than other types of medical intervention. Many of the above characteristics also present therapeutic opportunities.[§§] As gene-transfer interventions become more familiar or safeguards more sophisticated, many of these factors will cease to present significant sources of uncertainty. And some of the issues described above are theoretical and might not ever emerge as major risks; for example, with the possible exception of the X-SCID study, transgenes have never demonstrated unexpected properties in human beings. The suggestion is instead that each of the above characteristics heightens the degree of uncertainty surrounding gene-transfer risk estimates – given current knowledge.

What are the implications of greater uncertainty? One answer is: there are none. According to this view, clinical experiments are, by definition, forays into the unknown. Deaths and adverse outcomes are a regrettable but inevitable by-product of any attempt to extend the frontier of biomedical knowledge. The response to extreme uncertainty should be the same as for familiarity: make the best risk estimate you can, and give it a go.

One reason this view seems unsatisfactory is that it elides the question of what constitutes a "best-risk estimate." For a well-characterized drug, a best-risk estimate will be considerably more predictive than for novel drugs. The above response merely displaces the question of

---

makes pre-clinical evaluation of the risk in animal experiments much more difficult, and sometimes perhaps impossible."[27]

‡‡ As of this writing, the role of adenoviral vectors in disadvantaging volunteers is statistically unproven, because the study was not designed to test a possible adverse relation, and was cut short because of early trends of non-efficacy. Again, possible adverse consequences were not predicted in previous animal studies: as one NIH researcher put it, "some monkeys have lied to us this time."[28]

§§ "Smart," conditionally replicating vectors present the possibility of a drug that can have highly specific and selective effects. Intervening within genetic pathways is precisely what makes gene transfer so attractive a strategy. The interaction of a gene-transfer agent's subcomponents might be exploited. For example, viral vectors might be used to amplify an immunological reaction against antigens presented on the surface of a patient's tumors. Last, continuous, low-level exposure to a transgene product could have advantages for diseases that involve enzyme deficiencies. It may have other advantages as well. For example, there is some evidence that continuous, low-level exposure to the clotting proteins factor

how to manage uncertainty with an equally intractable question: how much safety data are enough?

Another response would begin by taking a more careful look at the product formula. Typically, the probabilistic component of the product formula gets glossed as a single, determinate probability. But consider the following scenarios. Investigator A wishes to pursue a clinical trial of a very well-characterized drug. Based on a large body of clinical evidence, the investigator estimates that for every 100 volunteers, one will experience a life-threatening liver toxicity. Investigator B has developed a novel gene-transfer approach that she believes is much safer. By definition, she is unable to access any body of clinical experience to estimate liver toxicity. She has, however, carried out an extensive program of preclinical toxicity testing in two different non-human animals; using a series of extrapolations from these data, she estimates a human liver toxicity risk of 1 in 500. Which is the riskier clinical trial?

The simplistic answer would compare probabilities and declare the second safer by a factor of five. What this estimate misses, however, is recognition that risk encompasses confidence as well. For drug A, the investigator is not absolutely certain that the risk is 1 in 100. He is 95 percent certain, however, that the risk is no greater than 1.2 in 100. Investigator B, however, has based her risk calculation on a smaller sample of animals, and her estimate incorporates a series of assumptions about the similarity between her test species and human beings. When asked to estimate the level of risk that she believes to represent the outer edge of 95 percent confidence, she answers 1 in 50. Now which trial appears to be the safer one?

I do not wish to suggest that there are simple ways of adjusting risk estimates using "uncertainty factors" based on animal data. I also do not want to fixate on the 95 percent confidence figure.[¶¶] Instead, the point is that uncertainty matters with respect to estimating risk, and that highly indeterminate studies can involve greater risk than determinate ones even if a "best guess" suggests that the former are safer.

Which risk matters with respect to the ethical analysis of risk in a first-in-human trial: the "best estimate", or the upper bound estimate? For two reasons, I would argue that the "upper-bound" estimate seems more appropriate. First, it is consistent with the mandate of minimizing risk in research.[39] Early-phase trials, as will be argued in the next chapter, primarily aim at testing safety. The risks of the intervention are therefore not generally justified by direct medical benefits for volunteers, particularly in first-in-human studies, which involve subtherapeutic doses.[†††] If a clinical investigator wishes to ask a volunteer to take a non-therapeutic risk, non-malificence would require a reasonable level of confidence that a given harm will not occur. Second, the "upper-bound" approach is more consistent with current policy on risk. Within the realm of first-in-human studies, for example, the FDA requires that sponsors of first-in-human drug trials involving healthy volunteers calculate phase 1 starting doses on the basis of "no adverse effect levels" (NOAELs), which refer to the highest dose at which a drug can be given without causing toxicity. Once a dose has been calculated, the FDA typically reduces starting doses by a

---

XIII prevents or attenuates the development of immunological reactions against the protein in hemophilia patients.[36,37]

¶¶ In some sense, the types of uncertainty in first-in-human trials are unmeasurable, because variation between model species and humans is not known beforehand. Such "manufactured" and "unmeasurable" risks are the types of entities that "risk society" theorists have written about.[38]

††† There are, of course, exceptions here. But the claim represents a general rule and has advantages in terms of operational simplicity.

factor of ten if there are significant uncertainties about the projection of animal data to human beings[40] (European regulators are considering using a somewhat more conservative method, the "Minimal Anticipated Biological Effect Level" (MABEL) for determining dose in first-in-human trials; when MABELs and NOAELs are different, the lower value should be used).[41] The Environmental Protection Agency (EPA) uses an approach that is still more precautionary. For example, it sets policy on exposure to air pollutants on the basis of "upper-bound" estimates of cancer risk.[42] Both approaches involve procedures that respond to uncertainty by defining risks conservatively.

Uncertainty of principles: justifying risk

How should investigators respond to uncertainty? There is virtually no limit to the ways that risks can be minimized in first-in-human gene-transfer trials, and I will keep my comments on risk minimization brief by describing four broad strategies. The first is choice of intervention strategies. Vectors can be altered to dampen or eliminate certain risks. Thus, the substitution of first-generation adenoviruses with highly deleted adenoviral vectors reduces some of the risk of immunogenicity;[43] the use of self-inactivating retroviruses in ex vivo protocols may lower the risk that vectors will activate oncogenes proximate to their insertion sites.[44]

A second approach to risk minimization is cautious selection of subjects. This was discussed in the previous chapter; in general, protocols that are subject to major indeterminacy should generally be reserved for treatment-refractory populations. Where trials are performed in medically stable or healthy volunteers, the protocol should exclude volunteers at greater risk for adverse events. For example, hemophilia gene-transfer protocols reduce the risk of generating immune responses against transgenes by excluding volunteers who show evidence of neutralizing antibodies against intravenously administered coagulant factor.[45]

A third broad strategy is trial design. For example, a disastrous trial involving the monoclonal antibody TGN1412[‡‡‡] drew attention to the need for dosing volunteers in sequence, rather than in parallel, where interventions exploit novel biological pathways.

Fourth, greater uncertainty generally calls for closer monitoring of study volunteers. At least two broad approaches exist. The first is through monitoring volunteers for subclinical anomalies relating to theoretical risks.[§§§] Protocols routinely collect tissue samples, and monitor for certain subclinical effects. However, they often apply a very limited panel of tests unless a clinical adverse event would prompt them to look further. For example, ex vivo protocols involving hematopoietic stem cells routinely test blood samples for clonality, which would indicate that some cells have a significant proliferative advantage that could develop into a cancer. Similarly, investigators might monitor immune status and cytokine profiles of volunteers if a vector is postulated to be immunogenic. The second monitoring strategy involves follow-up of volunteers for latent adverse events (this is discussed below).

[‡‡‡] In this study, six "healthy volunteers" participating in a phase 1 trial were hospitalized after receiving a monoclonal antibody designed to modulate immune responses in patients with B-cell chronic lymphocytic leukemia and rheumatoid arthritis. Two patients required "intensive organ support."[46,47]

[§§§] A similar recommendation was recently offered for monitoring changes in cognition, mood, and behavior arising from cell-based interventions in the brain. The authors urge neuropsychological testing before and during trials; they also urge assessment of matched controls for comparison.[48]

With respect to benefit, the "upper-bound" risk estimate approach defended above, together with the position that would question the use of direct medical benefits to justify trial risks in first-in-human studies (see Chapter 5), would seem to suggest that trials involving highly novel interventions should be held to a more stringent standard of scientific value.[¶¶¶] And given that such studies are primarily directed towards characterizing the safety of interventions, a good portion of this scientific value will be captured in characterizing risk and dispelling uncertainty about safety.

Many investigators would likely object to the suggestion that gene-transfer trials – or any novel intervention trial – be held to a more stringent standard of value. They might argue that this deters gene-transfer research, and imposes an unfair burden on the field, and perhaps even on expectant patients. They might further argue that the logic of my argument is inconsistent with the assertion of a more stringent standard: if the FDA (or IRBs) are already applying safety factors or using upper-bound risk estimates, then the risks of a first-in-human study of gene transfer should already be equivalent to the risks of a phase 1 trial of a well-characterized agent. These positions have some merit. Nevertheless, I think there are two reasons to reject them. The first concerns benefit. When investigator A began his trial, he had a significantly higher level of confidence that the results would translate into a clinical product, perhaps because the drug was very similar to another drug that is already licensed. Investigator B recognized that translating her novel product smoothly into a clinical application was a long shot. The "lower bounds" of utility probability for the latter trial is quite a bit lower than that for the novel product if we define utility strictly in terms of leading to phase 2 trials. As such, the expected value of the novel protocol is lower, as is its benefit–risk ratio.

The second argument for a stringent value standard concerns opportunity cost: because of the uncertainties surrounding them, novel intervention studies offer unique opportunity for advancing medical knowledge. To pursue such trials as if they were nondescript drug studies would fail to actualize their potential to advance science. This leads to a second dimension of the ethical encounter with uncertainty: activities occurring outside the specific protocol.

Expecting the unexpected: organizing uncertainty

As hinted at in the previous chapter, risk and uncertainty are, in a sense, "organized." That is, risks emerge as a consequence of various social and organizational practices that transcend individual actors or protocols. And it is at the level of organizations – in the case of gene transfer, the RAC, FDA, IRBs, a research field, and individual teams – that risks are conceptualized and managed.[49] This organizational dimension raises challenging moral questions in part because it disperses agency and responsibility for harms. It is also not the traditional locus for ethical inquiry, at least not within research ethics.[****]

¶¶¶ I note here an important proviso. Not all research ethicists accept the simple utilitarian calculus that balances out risk to subjects against benefit to society. I agree that there is something problematic here, in part because it draws no theoretical limit on risks that can be presented to volunteers. But the position is so firmly entrenched within the standard practice of research ethics, and this is not an appropriate venue to take on this foundational question.

**** The organizational approach to risk has been applied to medicine in other realms, however. A 1999 Institute of Medicine report, *To Err Is Human*, drew attention to organizational and system dimensions of medical error.[50,51]

The field of gene transfer has sometimes been seen as emblematic of a series of risk-abetting organizational shortcomings, many of which are described in the previous chapter: the incentive structure in biomedical research favors risk-taking and haste over caution, the prestige and financial rewards of early entry attract risk-prone "clinical champions" and biotechnology firms, and researchers often develop attachments to particular approaches that can, in the words of one panel of gene-transfer researchers, "cloud decisions."[52] Yet in reaction to major setbacks like those described in this and the previous chapter, the field has also built unique and sophisticated compensatory structures that indicate how other novel fields might "tame uncertainty." (Box 4.1)

Box 4.1 Shaping uncertainties in gene transfer

Over the years, gene-transfer researchers and regulators have evolved a number of formal and informal practices that shape and reduce uncertainties. The first is centralized and transparent review. Through the National Institutes of Health (NIH), the Recombinant DNA Advisory Committee (RAC) provides a mechanism for integrating expertise under indeterminate circumstances. It also provides an additional layer of safety and ethical oversight. The NIH also requires that research teams affiliated with publicly funded institutions pursuing recombinant DNA research report all serious, unexpected adverse events to the agency.[53] These reports are discussed in a public forum and posted in abbreviated form on the World Wide Web.[54] An additional RAC function is the organizing of "gene therapy policy conferences" and "safety symposia." These one-day meetings have typically brought together leading figures in gene-transfer research to analyze data and discuss the risk surrounding particular vectors or intervention strategies. The objectives of such conferences include "interagency communication and collaboration" and "concentrated expert discussion."[55,56]

The National Gene Vector Laboratories (NGVL), established in 1995 to provide clinical-grade vector for publicly funded phase 1 gene-transfer trials, is a second example. From 2001 to 2008, the NGVL funded or provided facilities for toxicology testing, and maintained a public database of vector toxicology studies; it also provides tissue archiving services.[57] The latter two functions were transferred to the National Gene Vector Biorepository after the NIH discontinued funding for NGVL in 2008.[58,59]

Since 1999, the FDA has asked sponsors to monitor the health of volunteers receiving retroviral vectors for fifteen years. Months before lymphoproliferative disorders were detected in the X-SCID study, an advisory body within the agency recommended a more comprehensive, fifteen-year follow-up policy. The agency's final guidance document was issued in 2006; it asked research teams to test tissue samples for malignancies in protocols involving integrating vectors.[60] The FDA also issued guidance encouraging sponsors of early-phase gene-transfer trials to establish data safety monitoring boards (DSMBs).[61,62] These bodies review safety data in order to identify specific trends. They are also charged – perhaps paradoxically – with making sure trials meet their recruitment and scientific goals.

The research community has also organized a series of workshops and efforts aimed at integrating current safety knowledge, establishing best practices, and defining a research agenda. One example is a working group established by the North American scientific society American Society of Gene Therapy (ASGT)[††††] to review animal and clinical data on the safety of retrovirus gene transfer. Their report identified numerous desiderata for future studies.[63] Other examples include workshops to discuss "obstacles preventing gene therapy from realizing its

[††††] Disclosure: I am a member of this organization and serve on its ethics committee, though I have not had any dealings with organizing safety workshops.

Box 4.1 *(cont).*

full potential,"[64,65] and several retreats run in conjunction with biomedical society meetings to assess current knowledge and research needs concerning the risks of integrating vectors.[‡‡‡‡]

Last are informal practices and networks within the gene-transfer research community. Marina Cavazzana-Calvo, for example, described the period after the first lymphoproliferative diagnoses as a "very difficult moment that allowed the scientific community to actively collaborate quickly and efficiently; numerous researchers offered reagents or services without asking for anything in return."[66] Indeed, the publication that first reported the disorders listed 35 authors from 21 different institutions located in six different countries.[67] Another example is the practice within new research fields of perennial grooming and culling of research data. The scholarly literature on gene transfer is awash in review articles. These help organize and systematize disparate and otherwise unlinked research findings, and help articulate unmet research needs.[§§§§]

First-in-human trials are hardly the first time human beings have employed risky and unpredictable technologies. Other examples include space shuttle launches, nuclear weaponry, and chemical manufacturing plants. All share several features. For one, they are complex and risks emerge at multiple junctures in their application. With gene transfer, manufacture, administration, investigator and patient behavior all involve risk. Second, risks are often difficult to model or predict: operation rarely proceeds exactly as planned, and worrisome signals are often hard to discern above the noise of "routine" error and malfunction (DSMBs, for example, face challenges in sifting adverse-event reports for trials involving morbid disorders like advanced cardiovascular disease). Third, the risky and unpredictable technologies described above are all rapidly evolving: components and software are quickly replaced with newer versions. In gene transfer, vectors or strategies might be outmoded before a trial has finished enrolling volunteers. Last, all must mediate tensions between imperatives for efficiency and a mandate for caution.

According to many social scientists, such complex and indeterminate hazards are best negotiated by institutions that possess the characteristics of "high-reliability organizations."[¶¶¶¶] The first characteristic is prioritization of safety over objectives like production. That is, an

‡‡‡‡ "Stem Cell Clonality and Genotoxicity Retreats." To date, three have been held. The first was in conjunction with the American Society of Hematology (AHA) Meeting in San Diego, 2003; the second was in conjunction with the ASGT meeting in St. Louis; the third was in 2006 (AHA, Orlando, FL).

§§§§ This impression is partially corroborated by a bibliometric study undertaken by my research group comparing the characteristics of publications that cite phase 1 conventional cancer studies with those publications that cite phase 1 gene-transfer studies. We found that gene-transfer trials were far more likely to receive many citations in reviews than conventional-agent trials.

¶¶¶¶ What follows is a description of "high-reliability organizations," which is a theory of risk management that originated from a series of sociological studies of high-risk industries and technology. This theory is often contrasted with "normal accident theory" of Charles Perrow and colleagues, which views certain catastrophic failures as a virtual inevitability where technologies are complex, and their operation is "tightly coupled."[68,69] While this is not the appropriate venue for an explication of these competing theories, I am defending my preference for the former in this context on two grounds. First, gene-transfer clinical trials are not the types of tightly coupled scenarios Perrow and colleagues seem to have in mind with normal accidents. Specifically, the process by which protocols are designed and reviewed is a notoriously slow and methodical one with considerable buffer for iterative changes, deliberation, adjustment, and delay. Second, while high-reliability organization theory occasionally descends into what appears (to this outsider) like a hokey management-speak (terms like "heedfulness" and "sense-making" permeate

organization must be prepared to make major sacrifices in order to ensure safety. The second is a culture of reliability and safety: managers and lower-tier personnel must feel comfortable expressing safety concerns, disclosing human error, and reporting "near misses." The third is redundancy in safety mechanisms. In general, the greater redundancy, the more likely it is that error and risky deviance will be corrected within an organization. In particular, serial checks are more likely to be precautionary than parallel checks, because the former provide fewer opportunities for a concern raised in one forum to be overruled by another.[70] The last is a commitment to "organizational learning," in which information is pooled and reviewed periodically, and practices updated in response to anomalies, near misses, and catastrophes.

To varying degrees, many of these organizational features are recapitulated in the policies and practices of gene transfer. Redundancy provides a good example. Gene-transfer researchers often bemoan the baroque oversight system that has befallen their field.***** A typical, NIH-funded protocol might undergo review by (1) a granting agency, (2) a protocol review committee, (3) a data safety monitoring board, (4) an institutional biosafety committee, (5) the RAC, (6) the FDA, and (7) an institutional review board. At any point in the process, a protocol can get kicked back to a body that has already completed its review. Though such redundancy represents a major nuisance and expense, it serves a critical function in maintaining safety.[72]

Table 4.1 lists numerous voluntary and policy-imposed systems for organizational learning in gene transfer. Still, one feature that distinguishes gene-transfer trials from, say, space shuttle launches is their protean organization. Shuttle risks are managed within a highly structured and hierarchical bureaucracy at NASA. The field of gene transfer, for a variety of reasons, barely coheres, and the links that bind researchers to each other and their field are less stable. This organizational fluidity may represent an important asset for cutting-edge research areas like gene transfer, which flourish by granting researchers wide latitude to tackle problems in innovative ways. But it also presents liabilities.

The role of standards and other trial design features in organizing uncertainty

Translational research – particularly gene transfer throughout the 1990s – often has a tournament-like incentive structure, with prestige and reward flowing disproportionately to investigators who first attempt or achieve success with an intervention. Research groups therefore frequently act as isolated and uncoordinated units. Commercial pressures further impede the development of risk knowledge: biotechnology companies are often more financially precarious than large pharmaceutical firms. While the penalties of an unexpected problem might mean their demise, the culture of venture capital is not one that encourages plodding caution. Moreover, such companies are often reluctant to share their preclinical data for fear that competitors will free-ride off their research investments.

---

this literature), the policy recommendations that these theorists advance seem sensible and consistent with many reforms already implemented without the guidance of theory.

***** Referring to the Beatles' *Helter Skelter*, one researcher likened it to a helter skelter ride: "When you get to the bottom you go back to the top of the slide/and you stop and you turn/and you go for a ride/then you get to the bottom/then you see me again."[71] It is not uncommon for a gene-transfer investigator to hire a full-time regulatory affairs officer for their own research laboratory.

Table 4.1 Controlling uncertainty by organizing learning in novel intervention trials

Level	Practice
Individual practices	Full publication of preclinical toxicology studies
	Performing toxicology studies in animals that recapitulate essential characteristics of subject (e.g. age, sensitivities and disease states, etc.)
	Testing for delayed adverse events in preclinical models
	Monitoring subclinical toxicities
	Long-term follow-up of volunteers
	Autopsy, analysis of banked tissues
Collective practices	Establishing trial design standards for integration of results into meta-analysis
	Structures promoting analysis of pooled data
	Standards for measuring dose and toxicities
	Regular meetings and working groups to review current risk knowledge and research needs

Committees convened after the University of Pennsylvania and Paris-Necker episodes point to three specific impediments to learning in gene transfer. First, translational research tends to revolve around showing efficacy rather than risk. For example, an ASGT committee observed that most published studies were "designed primarily to evaluate efficacy of gene transfer, gene expression, and the effects on disease phenotype" rather than safety,[63] and researchers often cut their preclinical studies too short to assess latent toxicities.[63] An NIH panel noted that trial teams often missed opportunities to "confirm or reject hypotheses about vector safety" by "not conducting autopsies on deceased subjects,"[73] faulted investigators for not collecting "information about the critical element of vector distribution in research participants ... [or] how the route [of administration] may affect the eventual distribution of each of the vectors," and noted the deficient use of proper controls in clinical and preclinical experiments. The same panel also recommended that investigators "systematically monitor" the immune status of research subjects and collect cytokine profiles.[73]

Second, certain preconditions for organizational learning are not in place. Following the OTCD death, researchers and regulators wanting to integrate safety data across multiple trials were stymied by variation in how researchers performed their clinical and animal studies.[†††††52,64,75] Take the seemingly straightforward concept of measuring dose: as described in Chapter 3, gene-transfer agents are composite entities, and a given preparation will contain a mixture of functional vectors that are capable of transfecting human tissue, as well as incompletely formed vector particles. From an immunological standpoint, both might trigger innate immunity. But from the standpoint of the gene-transfer objectives, only the former is capable of causing transgene expression. Which is more appropriate to use when reporting vector dose? And, more specifically, which assay among the numerous ones available is most appropriate? As one report stated: "The lack of uniformly recognized standards for measuring vector concentration or potency from one gene-transfer research group to

[†††††] In the words of sociologists Stefan Timmermans and Marc Berg, "standards aim at making actions comparable over time and space."[74]

another ... complicates meaningful comparisons of the raw data across clinical trials. Vector titer standards, model Ad[enoviral] vector particle measurement assays, and suitable reagents and reference standards need to be available for use among all laboratories involved in these efforts. Standardized assays should also be developed for use with vectors that are based on viruses other than adenovirus and with nonviral vectors."[73] The first adenoviral reference standard became available in 2005.[76] Its utility for measuring other adenovirus serotype doses, however, is unclear.

As of this writing, eleven years have passed since the first AAV studies in human beings. To date, there have not been any major mishaps that are clearly attributable to the vector though several near misses have occurred in animal[77] and clinical studies.[10,78] In 2002, the FDA urged the AAV community to establish a standard, and to that end, researchers formed a working group. The process is near completion for one serotype,[79] AAV2, which means that had a major adverse event occurred in an AAV trial until now, forensic studies would have encountered obstacles like those after the OTC study. The reasons for the delay are complex, and involve resources and shifting science. But according to three AAV researchers involved in the effort, "The primary hurdle is motivation. There is a lack of urgency in the AAV community to develop an AAV RSS [reference standard stock]. This is in contrast to the high-quality adenoviral reference standard that was developed out of necessity following the unfortunate death of a patient during an adenovirus gene therapy trial."[80]

Dosing standards are but one example. Another is study design: "If proper controls were used more widely, and if gene transfer experiments in general were more uniformly designed, it might also become possible to combine vector safety-related observations from many protocols for the purpose of conducting a trend analysis."[73] Another is measurement of immunological status:[‡‡‡‡‡] "establishment of a standardized approach to gauge immune competence prior to administration of a gene transfer product would be very useful in assessing the pharmacodynamic effects of gene-transfer products on the immune system."[73] Another is tissue specimen archiving: "Protocols for sampling, collection, processing and storage (pre and post therapy) should be standardized."[81]

Third, even where data are amenable to synthesis, few formal structures have been established for doing so. One working group stated "there is no central place one can go to for a well-analyzed summary of these data ... for example, a rigorous 'horizontal analysis' of the data from adenovirus vector trial would be invaluable to current or pending trials. Again, this is an obvious area of need that has been woefully ignored."[64] Two years later, the same group continued to lament the abundance of "incompletely analyzed data 'out there' from which much information can be gathered."[52] Similarly, a working group reviewing the NGVL program concluded that its toxicology database "could be used more widely."[82] The database contains only eighteen reports (as of October 18, 2007) and, according to director Kenneth Cornetta, "Everyone [NGVL] have asked so far, whether funded by NGVL or not, has agreed to submit" toxicology data to the database. But he also acknowledged that getting toxicology data into the public domain is "a culture thing we need to work on." (K. Cornetta, personal communications, 2008, 2007.) And according to personnel who run the program, only 40 passwords have been issued to access the database (L. Matheson, personal communication, 2007.)

‡‡‡‡‡ For another example, see CABO II 487–8: "the first issue on the Cabo II 'to do' list is promoting consensus on what assays will be meaningful, and how to standardize these assays to allow interstudy comparisons."[52]

Recommendations for taming of uncertainty

Numerous other collective practices and reforms might further advance the control of uncertainty (summarized in Table 4.1). These divide into two categories: those that improve the assessment and characterization of risk across the life cycle of a protocol (vertical reforms), and those that enable assessment of risk across multiple studies (horizontal reforms).

Modifications across the life cycle of protocols

Vertical reforms might begin with the publication of preclinical toxicological studies. Private drug developers typically protect such studies as trade secrets, and article 39.3 of the Agreement on Trade-Related Intellectual Property Rights (TRIPS) requires that drug regulatory authorities in signatory states maintain preclinical data in strict confidence.[83] However compelling the commercial rationale, the ethical reasons for divulging toxicological data seem overwhelming. First and foremost, nonpublication blocks one path for organizational learning, and in so doing, exposes patients and human subjects to needless risk. Second, nonpublication represents a terrible waste of animals. Third, where preclinical data are collected in relation to work emerging from academic medical centers, nonpublication frustrates the university mission of knowledge creation and public service.

A policy that compels drug developers to publish preclinical toxicological studies might strike some readers as impossibly idealistic. Until recently, however, Japan required full disclosure of preclinical studies as a condition for licensure,[84] and countries like Brazil, Israel, and Canada maintain far weaker data exclusivity policies than the USA.[85] Moreover, the policy would be analogous to the requirement of elite medical journals that researchers deposit genetic-sequence or protein-structure data in public databases as a condition for publication.[§§§§§]

Several recommendations can also be offered for the design of preclinical safety studies. Such studies should, as much as possible, emulate characteristics of the clinical setting.[63] Thus, for example, the age and health status of animal models should match those of the anticipated subject population. Preclinical studies should also run sufficiently long to detect latent effects like transformation or autoimmunity induction. Finally, as per the recommendations above, preclinical studies should measure various toxicity surrogates and, where possible, use and report on standardized assays.

At the opposite end of the trial are mechanisms for detecting delayed adverse events. As noted previously, the FDA has issued guidelines encouraging researchers to follow gene-transfer recipients for fifteen years. Since 2000, the Gene Therapy Advisory Committee in the United Kingdom has promoted life time follow-up of gene-transfer recipients: upon consent, the health records of gene-transfer recipients – and their offspring – are “flagged” such that the GTAC can contact general practitioners to gather information about possible adverse outcomes. However, both policies have important limitations. The FDA’s is nonbinding and pertains only to integrating vectors, specifically, retroviral and lentiviral vectors. Vectors that are maintained episomally (that is, not integrated within a chromosome)

[§§§§§] For example, *Molecular Therapy* – the publication of the American Society of Gene Therapy – requires submission of data to “appropriate database(s)”. *Lancet* and *Science* also require authors to include accession numbers in submitted manuscripts.[86,87]

over long periods or that integrate at low frequency are thus excluded.[88] ¶¶¶¶¶ Given current debate about whether AAV vectors can cause cancer,[77] such exemptions seem regrettable.****** As for the UK program, a 2005 report found that "flagging returns were much below the expected rate, there [sic] was limited uptake by investigators and/or patients. This suggested low confidence in the project by trialists and/or patients."[90] The flagging project was discontinued in 2007.

Some question the value of long-term follow-up, pointing out that sample sizes, disease morbidities, and pretreatment greatly complicate the assessment of delayed adverse outcomes. Others note that long-term follow-up demands considerable resources, and is particularly difficult in settings like the USA, where no single authority tracks the health records of patients.[91] Both seem solvable with sufficient commitment of energy and resources. For example, problems related to sample size or the confounding effects of natural disease course might be partly addressed by building a database of matched historical controls†††††† and by pooling data across multiple studies. Moreover, the objective of such a database is to identify issues that warrant hypothesis-driven safety testing. The problem of resources is not trivial: according to the FDA, annual costs for long-term follow-up can run between $1500 and $5000 per patient.[89] Still, the problem is one of priority on the part of funding agencies, investigators, and sponsors.‡‡‡‡‡‡

A related reform would be a more concerted effort to collect information from autopsies. The NIH and FDA encourage investigators to perform autopsies on deceased gene-transfer volunteers. However, many investigators report difficulty obtaining consent from surviving family members. Whether these difficulties arise because trialists are reluctant to solicit consent, or because bereaved family members often decline, is unclear. Empirical studies on autopsy consent provide some reassurance that, however sensitive the discussions, bereaved family members are generally receptive to solicitations for post-mortem examinations.¶¶¶¶¶¶ At the very least, policy on autopsy would have a sounder basis if investigators publicly reported the frequency with which the families of volunteers were approached for autopsy, and how often they were refused.

¶¶¶¶¶ Another limitation of the FDA guidance is that it takes a "vector"-centered view of long-term follow-up needs. Though vector properties are likely to be the most significant physical determinant of trial risks, latent adverse events can also occur with benign vectors. For example, many critical limb ischemia trials involve expression from vectors that are only transiently retained in the body. Still, such protocols are designed to induce permanent vascular changes, which in principle could have adverse outcomes by stimulating tumor growth or disorganized blood vessels. Both would be difficult to rule out without follow-up of subjects.

****** The position is consistent with that of the Biological Response and Modifiers Advisory Committee, BRMAC, which formerly advised the FDA about gene transfer policies: "The BRMAC deliberated on the risks associated with RNA- and DNA-based vectors and the various factors that contributed to those risks. They noted that all gene-therapy vector systems carry sufficient concerns about long-term side effects to warrant some level of long-term follow-up".[89]

†††††† Duggan *et al.* also recommend matched controls as a way of monitoring adverse events.[48]

‡‡‡‡‡‡ One example of innovative techniques to enhance follow-up is provided by: Sprague *et al.*[92]

¶¶¶¶¶¶ The authors of one report emphasize the importance of describing how, precisely, the autopsy might contribute knowledge gathered in the context of a clinical trial.[93] Another recent study indicating bereaved family member receptivity to autopsy – outside the context of a clinical trial – is provided by Rankin *et al.*[94]

Modifications enabling risk monitoring across multiple protocols

In addition to the horizontal reforms suggested by official gene-transfer bodies (see section "The role of standards and other trial design features in organizing uncertainty", above), several others might be offered. One would be the design of phase 1 studies for planned integration into a meta-analysis. Early-phase clinical studies are substantially underpowered for detecting all but the most extreme toxicities. By standardizing certain design features (e.g. dose measurement, immunological assays, other surrogate safety endpoints, toxicity grading, adverse-event reporting,******* etc.), researchers might enable the detection and characterization of less-common toxicities.[96]

Another reform would be the creation of structures and resources that would promote horizontal analysis. One of the major challenges in any novel field is the problem of anticipating the unforeseeable. As one gene-transfer researcher put it when asked about the monitoring of volunteers for unexpected adverse events, "how can you look for what you don't know what to look for?"(S. H.-B. Abina, personal communication, 2007) One way of addressing this problem would be to create a centralized databank of tissue samples collected from volunteers before and after receiving gene transfer. Provided adequate informed consent and a procedure to strip specimens of personal identifiers, such a databank might prove useful for testing specific hypotheses about toxicities. The NGVB currently maintains a tissue repository of gene-transfer specimens. However, the services they offer have two major drawbacks: they extend only to retroviral protocols, and the databank is intended purely as an archive for individual research teams rather than as a resource for independent investigations.[97]According to Kenneth Cornetta, researchers would generally oppose opening the archive to independent studies (K. Cornetta, personal communications, 2007.)

Last, preclinical data might be pooled regularly to enable a sort of translational "pharmacovigilance." Meta-analysis of preclinical data is exceedingly rare, but might provide a more comprehensive measure of risk.[63] Again, this would require a core set of standards for measuring toxicities, dosages, and vector characteristics, as well as a willingness to publish toxicity studies.

On whose shoulders does the moral responsibility for implementing these recommendations fall? The question presents particular challenges in the context of so loosely organized a scientific enterprise. Still, several parties to gene-transfer research seem obvious candidates. Professional societies like the American Society of Gene Therapy (ASGT) might play a more active role in helping to devise standards, censuring investigators whose practices deviate from safety norms, and stimulating broad-ranging discussions about ethics among investigators. Funding agencies like the NIH might be more proactive in sponsoring integrative research, or standardizing the design and reporting of preclinical and clinical studies. The private sector might establish research consortia centering on common safety concerns. Last, many leading journals have endorsed policies aimed at enhancing the reporting of randomized controlled trials.[98] Leading gene-transfer journals should establish analogous structures for the reporting of preclinical and clinical studies.

******* One frustrating problem with adverse-event reporting is the tendency for teams to report events in ways that have limited utility. For example, adverse events are often reported on a per-procedure basis, raising questions of how frequently they occur on a per-patient basis. Some protocols report adverse events without specifying which patient they occur in, leaving it unclearwhether a cluster of adverse events occurred in a single patient or whether it was spread among different patients. Another problem is a tendency to not specify which adverse events were monitored but undetected. For an overview of adverse-event reporting problems in the context of a specific procedure (deep brain stimulation), and some constructive suggestions for overcoming them, see Videnovic & Metman.[95]

Conclusions

The position that risk and uncertainty are, in some sense, manageable is at odds with the views of some. In an analysis of the disastrous TGN1412 phase 1 study, two commentators argued: "Eventually, unless we are no longer subject to the law of averages, someone will experience these rare events. This is unfortunate, but unavoidable if we seek to improve medical care. The only alternative is to forego progress against the terrible diseases for which we still lack cures and effective treatments."[99]

Trials *do* venture into unknown territory, and some risks resist control. Still, one wonders: why is it that some industries and organizations are remarkably adept at managing risk while others are less successful? If accidents are inevitable and catastrophe unavoidable, why is it that the risk of airline fatalities declined from 1 person in 9.2 million in the 1990s to 1 in 22.8 million between 2000 and 2005?[100]

The position taken in this chapter is less fatalistic. Biological uncertainty creates ethical demands on researchers and research fields. However profound and complex, uncertainty is organized and amenable to further organization. Notwithstanding an incentive structure that at times seems to work against caution and organizational learning, gene transfer has taken exemplary steps to manage uncertainties. But the field's taming of uncertainty remains a work in progress.

Volunteers who enter early-phase trials of novel agents might be thought of as rescue workers in the aftermath of an earthquake, dispatched to a large, unfamiliar building at peril to themselves in order to gather intelligence for a rescue operation. Sending more than the minimally required rescue workers to a specific region of the building, failing to synchronize the workers' watches or use common plans to describe various positions in the building, or not pooling gathered intelligence would expose rescue workers and earthquake survivors to needless risk and, in so doing, demean their value. The same seems true for volunteers sent into trials that are not optimally executed to maximize their yield of safety information.

There is another way to interpret the "law of averages" when it comes to trial safety. If a system is geared towards ensuring, say, 95 percent confidence that a trial-induced death will not occur, for every trial where a volunteer dies, another 19 studies will have applied safety mechanisms that seem, from the standpoint of the investigator at least, gratuitous, numbing, and bureaucratic. The challenge for cutting-edge research fields like gene transfer is, in part, to make the boring, standard, and custodial seem glamorous.

References

1. Ellis R. We felt Rhys dying in our arms. Only pioneering gene therapy could save our 'baby in the bubble'; The poignant diary of the parents whose son has made medical history. *The Mail on Sunday* 2002 April 14.
2. Hacein-Bey-Abina S, Garrigue A, Wang GP, Soulier J, Lim A, Morillon E, *et al.* Insertional oncogenesis in 4 patients after retrovirus-mediated gene therapy of SCID-X1. *J Clin Invest* 2008; **118**(9): 3132–42.
3. Orkin SH, Motulsky AG. *Report and Recommendations of the Panel to Assess the NIH Investment in Research on Gene Therapy.* National Institutes of Health, Office of Biotechnology Activities, 1995.
4. Woods NB, Bottero V, Schmidt M, von Kalle C, Verma IM. Gene therapy: therapeutic gene causing lymphoma. *Nature* 2006; **440**(7088): 1123.
5. Thrasher AJ, Gaspar HB, Baum C, Modlich U, Schambach A, Candotti F, *et al.* Gene therapy: X-SCID transgene leukaemogenicity. *Nature* 2006; **443**(7109): E5–6; discussion E7.
6. Howe SJ, Mansour MR, Schwarzwaelder K, Bartholomae C, Hubank M, Kempski H, *et al.* Insertional mutagenesis combined with acquired somatic mutations causes leukemogenesis following gene therapy

of SCID-X1 patients. *J Clin Invest* 2008; **118**(9): 3143–50.

7. Crystal RG, McElvaney NG, Rosenfeld MA, Chu C-S, Mastrangeli A, Hay JG, *et al.* Administration of an adenovirus containing the human CFTR cDNA to the respiratory tract of individuals with cystic fibrosis. *Nat Genet* 1994; **8**(1): 42–51.
8. Raper SE, Chirmule N, Lee FS, Wivel NA, Bagg A, Gao G-p, *et al.* Fatal systemic inflammatory response syndrome in an ornithine transcarbamylase deficient patient following adenoviral gene transfer. *Mol Gen Metab* 2003; **80**(1–2): 148–58.
9. Recombinant DNA Advisory Committee (RAC), Minutes of Meeting, September 6–7 2001.
10. Manno CS, Pierce GF, Arruda VR, Glader B, Ragni M, Rasko JJ, *et al.* Successful transduction of liver in hemophilia by AAV-Factor IX and limitations imposed by the host immune response. *Nat Med* 2006; **12**(3): 342–7.
11. Horstmann E, McCabe MS, Grochow L, Yamamoto S, Rubinstein L, Budd T, *et al.* Risks and benefits of phase 1 oncology trials, 1991 through 2002. *N Engl J Med* 2005; **352**: 895–904.
12. Antoine C, Muller S, Cant A, Cavazzana-Calvo M, Veys P, Vossen J, *et al.* Long-term survival and transplantation of haemopoietic stem cells for immunodeficiencies: report of the European experience 1968–99. *Lancet* 2003; **361**(9357): 553–60.
13. Cavazzana-Calvo M, Fischer A. Gene therapy for severe combined immunodeficiency: are we there yet? *J Clin Invest* 2007; **117**(6): 1456–65.
14. Will E, Bailey J, Schuesler T, Modlich U, Balcik B, Burzynski B, *et al.* Importance of murine study design for testing toxicity of retroviral vectors in support of phase I trials. *Mol Ther* 2007; **15**(4): 782–91.
15. Stone R. Bacteriophage therapy. Stalin's forgotten cure. *Science* 2002; **298**: 728–31.
16. Steidler L, Hans W, Schotte L, Neirynck S, Obermeier F, Falk W, *et al.* Treatment of murine colitis by *Lactococcus lactis* secreting interleukin-10. *Science* 2000; **289**(5483): 1352–5.
17. Broeke AV, Burny A. Retroviral vector biosafety: lessons from sheep. *J Biomed Biotechnol* 2003; **2003**(1): 9–12.
18. Rasko J, O'Sullivan G, Ankeny R, eds. *The Ethics of Inheritable Genetic Modification: A Dividing Line?* Cambridge, Cambridge University Press, 2006; 27.
19. Ylä-Herttuala S. Adverse effects of gene therapy: Hero or villain? *Gene Ther* 2003; **10**: 193.
20. Brown MP, Topham DJ, Sangster MY, Zhao J, Flynn KJ, Surman SL, *et al.* Thymic lymphoproliferative disease after successful correction of CD40 ligand deficiency by gene transfer in mice. *Nat Med* 1998; **4**(11): 1253–60.
21. Kaplitt MG, Feigin A, Tang C, Fitzsimons HL, Mattis P, Lawlor PA, *et al.* Safety and tolerability of gene therapy with an adeno-associated virus (AAV) borne GAD gene for Parkinson's disease: an open label, phase I trial. *Lancet* 2007; **369**(9579): 2097–105.
22. Baum C, Dullmann J, Li Z, Fehse B, Meyer J, Williams DA, *et al.* Side effects of retroviral gene transfer into hematopoietic stem cells. *Blood* 2003; **101**(6): 2099–114.
23. Morris JC, Conerly M, Thomasson B, Storek J, Riddell SR, Kiem HP. Induction of cytotoxic T-lymphocyte responses to enhanced green and yellow fluorescent proteins after myeloablative conditioning. *Blood* 2004; **103**(2): 492–9.
24. Kay MA, High K. Gene therapy for the hemophilias. *Proc Natl Acad Sci USA* 1999; **96**(18): 9973–5.
25. Aguilar LK, Aguilar-Cordova E. Evolution of a gene therapy clinical trial. From bench to bedside and back. *J Neurooncol* 2003; **65**(3): 307–15.
26. Second Cabo Gene Therapy Working Group. Cabo II: Immunology and Gene Therapy. *Mol Ther* 2002; **5**: 486–491.
27. Expert Scientific Group on Phase One Clinical Trials (2006) Final Report. Secretary of State for Health The Stationery Office: Norwich, UK; 77.
28. Cohen J. AIDS research. Did Merck's failed HIV vaccine cause harm? *Science* 2007; **318**: 1048–9.
29. Bostanci A. Gene therapy. Blood test flags agent in death of Penn subject. *Science* 2002; **295**(5555): 604–5.
30. Langezaal I, Hoffmann S, Hartung T, Coecke S. Evaluation and prevalidation of an

immunotoxicity test based on human whole-blood cytokine release. *Altern Lab Anim* 2002; **30**(6): 581–95.

31. Haley PJ. Species differences in the structure and function of the immune system. *Toxicology* 2003; **188**(1): 49–71.
32. Herzog RW. Immune responses to AAV capsid: are mice not humans after all? *Mol Ther* 2007; **15**(4): 649–50.
33. Li H, Murphy SL, Giles-Davis W, Edmonson S, Xiang Z, Li Y, *et al.* Pre-existing AAV capsid-specific CD8+ T cells are unable to eliminate AAV-transduced hepatocytes. *Mol Ther* 2007; **15**(4): 792–800.
34. Morral N, O'Neal WK, Rice K, Leland MM, Piedra PA, Aguilar-Cordova E, *et al.* Lethal toxicity, severe endothelial injury, and a threshold effect with high doses of an adenoviral vector in baboons. *Hum Gene Ther* 2002; **13**(1): 143–54.
35. Dobrzynski E, Herzog RW. Tolerance induction by viral in vivo gene transfer. *Clin Med Res* 2005; **3**: 234–40.
36. Cohn EF, Zhuo J, Kelly ME, Chao HJ. Efficient induction of immune tolerance to coagulation factor IX following direct intramuscular gene transfer. *J Thromb Haemost* 2007; **5**; 1227–36.
37. Beck U. *Risk Society: Towards a New Modernity*. New Delhi, Sage, 1992.
38. US Department of Health and Human Services. *Protection of Human Subjects: Criteria for IRB approval of research 45 CFR 46.111(a)(1)*. US Department of Health and Human Services, 2005; 1–12.
39. US Food and Drug Administration, Center for Biologics Evaluation and Research. *Guidance for Industry and Reviewers: Estimating the Safe Starting Dose in Clinical Trials for Therapeutics in Adult Healthy Volunteers*. US Department of Health and Human Services, 2002; 1–26.
40. Committee for Medicinal Products for Human Use (CHMP). *Guideline on Requirements for First-in-Man Clinical Trials for Potential High-Risk Medicinal Products*. Draft. London, CHMP, 2007.
41. US Environmental Protection Agency, Technical Transfer Network. Background on Risk Characterization. *1996 National-Scale Air Toxics Assessment*. US Environmental Protection Agency, 2007.
42. Alba R, Bosch A, Chillon M. Gutless adenovirus: last-generation adenovirus for gene therapy. *Gene Ther* 2005; **12** (Suppl. 1): S18–S27.
43. Yu S-F, Rüden TV, Kantoff PW, Garber C, Seiberg M, Rüther U, *et al.* Self-inactivating retroviral vectors designed for transfer of whole genes into mammalian cells. *Proc Natl Acad Sci USA* 1986; **83**(10): 3194–8.
44. Kimmelman J. Staunch protections: the ethics of haemophilia gene transfer research. *Haemophilia* 2008; **14**(1): 5–14.
45. Suntharalingam G, Perry MR, Ward S, Brett SJ, Castello-Cortes A, Brunner MD, *et al.* Cytokine storm in a phase 1 trial of the anti-CD28 monoclonal antibody TGN1412. *N Engl J Med* 2006; **355**(10): 1018–28.
46. Kenter MJ, Cohen AF. Establishing risk of human experimentation with drugs: lessons from TGN1412. *Lancet* 2006; **368**(9544): 1387–91.
47. Duggan PS, Siegel AW, Blass DM, Bok H, Coyle JT, Faden R, *et al.* Unintended changes in cognition, mood, and behavior arising from cell-based interventions for neurological conditions: ethical chanllenges. *Am J Bioeth* 2009; **9**(5): 31–6.
48. Hutter B, Power M. Organizational Encounters with Risk: An Introduction. In: Hutter B, Power M, eds. *Organizational Encounters with Risk*. New York, Cambridge University Press, 2005; 1–32.
49. Institute of Medicine. *To Err is Human: Building a Safer Health System, Crossing the Quality Chasm: The IOM Health Care Quality Initiative*. The National Academies, Washington, DC, 1999.
50. Sharpe VA, Promoting patient safety. An ethical basis for policy deliberation. *Hastings Cent Rep* 33; (2003) S3–S18.
51. Cabo II: Immunology and Gene Therapy. *Mol Ther* 2002; **5**: 486–91.
52. Recombinant DNA Advisory Committee (RAC). *Guidelines for Research Involving Recombinant DNA Molecules Appendix M-I-C-4*. National Institutes of Health, Office of Biotechnology Activities, 2000.

53. www4.od.nih.gov/oba/rac/documents1.htm
54. Recombinant DNA Advisory Committee (RAC). *Guidelines for Research Involving Recombinant DNA Molecules Appendix L.* National Institutes of Health, Office of Biotechnology Activities, 2000.
55. For a listing of conferences see: www4.od.nih.gov/oba/rac/meeting.html.
56. www.ngvl.org/include/pages/tox_db.php.
57. www.ngvl.org/ (Last accessed February 9, 2009).
58. ww.ngvbcc.org/Home.action (Last accessed February 9, 2009).
59. US Food and Drug Administration, Center for Biologics Evaluation and Research. Guidance for Industry. *Gene Therapy Trials: Observing Subjects for Delayed Adverse Events.* US Department of Health and Human Services, 2006.
60. US Food and Drug Administration, Center for Biologics Evaluation and Research. *Guidance for Clinical Trial Sponsors: On the Establishment and Operation of Clinical Trial Data Monitoring Committees.* US Department of Health and Human Services, 2002.
61. National Institutes of Health. *Further Guidance on a Data and Safety Monitoring for Phase I and Phase II Trials.* US Department of Health and Human Services, 2000. Available at: http://grants.nih.gov/grants/guide/notice-files/NOT-OD-00-038.html (Last accessed August 11, 2007).
62. Kohn DB, Sadelain M, Dunbar C, Bodine D, Kiem H-P, Candotti F, *et al.* American Society of Gene Therapy (ASGT) Ad Hoc Subcommittee on Retroviral-Mediated Gene Transfer to Hematopoietic Stem Cells. *Mol Ther* 2003; **8**(2): 180–7.
63. The First Cabo Gene Therapy Working Group. A prescription for gene therapy. *Mol Ther* 2000; **1**(5): 385–8.
64. Cabo II. Immunology and gene therapy. *Mol Ther* 2002; **5**(5): 486–91.
65. Cavazzana-Calvo M. In: Kimmelman J, ed. Paris; 2007.
66. Hacein-Bey-Abina S, Von Kalle C, Schmidt M, McCormack MP, Wulffraat N, Leboulch P, *et al.* LMO2-associated clonal T cell proliferation in two patients after gene therapy for SCID-X1. *Science* 2003; **302**(5644): 415–19.
67. Perrow C. *Normal Accidents: Living with High-Risk Technologies.* Princeton, Princeton University Press, 1999.
68. Sagan SA. *The Limits of Safety: Organizations, Accidents, and Nuclear Weapons.* Princeton, Princeton University Press, 1993.
69. Heimann L. *Acceptable Risks: Politics, Policy and Risky Technologies.* Ann Arbor, University of Michigan Press, 1997.
70. Lennon J, McCartney P. *"Helter Skelter" The White Album.* London, Abbey Road Studios, 1968.
71. Walters L. The oversight of human gene transfer research. *Kennedy Inst Ethics J* 2000; **10**(2): 171–4.
72. National Institutes of Health. Assessment of adenoviral vector safety and toxicity: report of the National Institutes of Health Recombinant DNA Advisory Committee. *Hum Gene Ther* 2002; **13**: 3–13.
73. Timmermans S, Berg M. Standardization in action: achieving local universality through medical protocols. *Social Studies of Science* 1997; **27**: 273–305.
74. Hutchins B, Sajjadi N, Seaver S, Shepherd A, Bauer SR, Simek S, *et al.* Working toward an adenoviral vector testing standard. *Mol Ther* 2000; **2**(6): 532–4.
75. The Williamsburg BioProcessing Foundation. *Adenovirus Reference Material.* Virginia Beach, VA, 2002.
76. Russell DW. AAV vectors, insertional mutagenesis, and cancer. *Mol Ther* 2007; **15**(10): 1740–3.
77. Kaiser J. Clinical trials. Gene transfer an unlikely contributor to patient's death. *Science* 2007; **318**(5856): 1535.
78. Moullier P, Snyder RO. International efforts for recombinant adeno-associated viral vector reference standards. *Mol Ther* 2008; **16**(7): 1185–8.
79. Flotte TR, Burd P, Snyder RO. *Utility of a Recombinant Adeno-Associated Viral Vector Reference Standard.* Williamsburg Bioprocessing Foundation; available through the "Reference Material Projects" link at: www.wilbio.com/. Last accessed October 23, 2007.
80. Recommendations of the Gene Therapy Advisory Committee/Committee on safety of medicines working party on

retroviruses. May 2005. *Hum Gene Ther* 2005; **16**(10): 1237–9.

81. National Institutes of Health. *Report and Recommendations of the Gene Therapy Working Group*. Bethesda, MD, 2005 November 1.
82. World Trade Organization. *Trade-Related Aspects of Intellectual Property Rights (TRIPS).* Marrakesh, World Trade Organization, 1994.
83. Rokuro H. Full disclosure of evidence in published papers to prevent drug disasters: Lessons from citizens' pharmacovigilence activities. *Clinical Evaluation* 2005; **32**: 65–98.[Note: this article is in Japanese, and I am basing this citation on the English-translated abstract].
84. International Federation of Pharmaceutical Manufacturers Associations. *Encouragement of New Clinical Drug Development: The Role of Data Exclusivity.* Geneva, Switzerland, International Federation of Pharmaceutical Manufacturers Associations, 2000.
85. www.sciencemag.org/about/authors/prep/gen_info.dtl#related.
86. Williams DA, Frederickson RM. *Instructions for Authors, Molecular Therapy*. www.nature.com/mt/author_instructions.html; accessed October 11, 2007.
87. US Food and Drug Administration, and Center for Biologics Evaluation and Research. *Guidance for Industry: Gene Therapy Trials – Observing Subjects for Delayed Adverse Events.* US Department of Health and Human Services, 2006.
88. US Food and Drug Administration. *Gene Therapy Patient Tracking System*. US Department of Health and Human Services, 2002.
89. Gene Therapy Advisory Committee. *Review of the GTAC/DH Flagging Project*: Note of the Discussion Held by GTAC on 21 February, 2007.
90. Nyberg K, Carter BJ, Chen T, Dunbar C, Flotte TR, Rose S, *et al.* Workshop on long-term follow-up of participants in human gene transfer research. *Mol Ther* 2004; **10**(6): 976–80.
91. Sprague S, *et al.* Limiting loss to follow-up in a multicenter randomized trial in orthopedic surgery. *Control Clin Trials* 2003; **24**: 719–25.
92. Snowdon C, Elbourne DR, Garcia J. Perinatal pathology in the context of a clinical trial: attitudes of bereaved parents. *Arch Dis Child Fetal Neonatal Ed* 2004; **89**: F208–11.
93. Rankin J, Wright C, Lind T. Cross sectional survey of parents' experience and views of the postmortem examination. *BMJ* 2002; **324**: 816–18.
94. Videnovic A, Metman LV. Deep brain stimulation for Parkinson's disease: prevalence of adverse events and need for standardized reporting. *Mov Disord* 2008; **23**(3): 343–9.
95. For a discussion of planned meta-analysis in clinical trials, see: Berlin JA, Colditz, GA. The role of meta-analysis in the regulatory process for foods, drugs, and devices. *JAMA* 1999; **281**: 830–4.
96. www.ngvl.org/.
97. Moher D, Schulz KF, Altman DG, for the CG. The CONSORT statement: Revised recommendations for improving the quality of reports of parallel-group randomized trials. *Ann Intern Med* 2001; **134**(8): 657–62.
98. Emanuel EJ, Miller FG. Money and distorted ethical judgments about research: ethical assessment of the TeGenero TGN1412 trial. *Am J Bioeth* 2007; **7**(2): 76–81.
99. Levin A. Airways in USA are the safest ever. *USA Today* 2006 June 29.

Succor or suckers? Benefit, risk, and the therapeutic misconception

Introduction

Patients with undiagnosed glioblastoma multiforme – an advanced form of brain cancer – typically arrive at the doctor's office complaining of headaches and weakness; a good many might also have experienced a seizure. Within a year, approximately half will be dead.

Progress against glioblastoma has been halting, and even modest improvements in treatment, like a median extension of survival of two and a half months,[1] are greeted as breakthroughs. And so it was that, in 1992, a 51-year-old woman, whose husband was a prominent lawyer and chair of the board of the San Diego Cancer Center, was diagnosed with glioblastoma. Conventional therapies were unsuccessful in controlling her disease, and the woman sought enrollment in a gene-transfer study being planned at the San Diego Cancer Center. The problem was that the protocol had not yet been reviewed by the RAC, which met every three months; the patient was expected to live no more than two.

A former client of the woman's husband knew Iowa's Senator Harkin, and offered to contact him to see whether political pressure might be brought to bear. Shortly thereafter, Harkin forwarded a letter to NIH head Bernadine Healy requesting that she waive normal RAC review, though the developer of the intervention, Ivor Royston, had previously stated that he was "not optimistic that this therapy will work for his patient."[2]

RAC members bristled at Healy's involvement; they furthermore considered the supporting preclinical data insufficient to justify the protocol. Still, ethicist and panel member Leroy Walters conceded that occasional relaxation of inclusion criteria was appropriate in order to "respond to … [a] particular patient without waiting until the next scheduled RAC meeting."[3]

Are first-in-human studies an act of compassion for volunteers? Many patients, investigators, and policy-makers believe they are.[4,5] This chapter examines the emergence of the position that translational clinical trials are a form of medical care. I next explore some of the ethical and scientific aspects of this view, and close with a series of recommendations aimed at navigating a genuine therapeutic ambiguity presented by phase 1 studies.

From opportunism to medical opportunity

The notion that early-phase trials present medical opportunities for desperately ill volunteers is relatively new. Owing to a series of highly publicized research abuses, policy-makers and members of the public have tended to regard clinical trial participation as burdensome. Over the past decade or two, this protectionist presumption has eroded, and trials have increasingly come to be viewed as offering patients access to cutting-edge medications, elite care, and closer medical observation.

Several factors account for this shift toward what one commentator called "inclusionism."[6] One is a relatively new form of disease-group engagement with clinical research. Until recently, research advocates largely deferred decisions about study methodology to scientists and clinicians allied with their cause. This changed with the advent of HIV activism. Many gay activists questioned whether a predominantly heterosexual biomedical establishment could be trusted to respond to a disease that was devastating their community.[7] With slogans like "red tape is killing us," HIV activists pressed for greater access to investigational drugs and revision of traditional trial methodologies.[8] Numerous other disease advocates have since emulated HIV activism.[9–11]

Another important factor driving inclusionist policies is a North American culture of technological optimism. Various historians have noted how US culture imbues technologies with almost supernatural qualities, a phenomenon some have called the "technological sublime."[12,13]* Nowhere is this more evident than in medicine, where the human genome sequence is described as "the book of life," research objectives as "holy grails," and dramatic disease-reversal in drug trials as a "Lazarus response."[17] Accordingly, cutting-edge research in fields like genetics is often regarded as a kind of deliverance.

One result is a series of inclusionist regulatory reforms for early-phase trials. For example, expanded access authorizes manufacturers to provide individual patients with life-threatening disorders access to experimental medications (free of charge) through single-patient and treatment investigational new drug (IND) applications.[18] The policy of accelerated approval permits the FDA to license new drugs on the basis of surrogate endpoints, like tumor shrinkage or cellular markers, rather than clinical data like survival. In addition the FDA is, at the time of writing, unrolling a series of policies to encourage first-in-human trials. One of these lowers manufacturing standards for drugs tested in phase 1 studies.[19]† Other new inclusionist policies include the NIH's public database of clinical trials, which was created to provide "easy access to information on clinical trials for a wide range of diseases and conditions."[20]

Clearly, many patients have come to regard early-phase research as a medical opportunity. As of 2004, over 42 000 patients had received experimental drugs through expanded access.[21] Survey after survey indicates that the vast majority of cancer patients entering phase 1 studies seek clinical improvement,[22–26] and patients or their guardians are often willing to go to extraordinary lengths to gain entry to translational studies. The *New York Times* recounts one story in which parents raised $200 000 to support a gene-transfer trial that might enroll their child with Canavan's disease. When the protocol was bottled up in ethics review, the parents enlisted Senator Tom Daschle and contemplated a $250 000 nudge for the host university.[27]

Many researchers also view trials as a vehicle for care.[28–30] For example, 42 percent of oncologists in one study indicated that their primary reason for enrolling subjects in clinical trials was ensuring access to "state of the art" care.[31] One survey of gene-transfer researchers involved in early-phase studies indicated that 46 percent thought that their trial would offer enrolling subjects medical benefits.[25] Some institutions, like the MD Anderson Cancer

---

* The term originates with historian Perry Miller in his unfinished meditation on the sublime in American culture, *The Life of the Mind in America: From the Revolution to the Civil War*.[14] Leo Marx subsequently develops the concept in *The Machine and the Garden: Technology and the Pastoral Ideal in America*. The term appears to have acquired a somewhat ambiguous meaning, with some authors (like Marx, Miller and Kasson) emphasizing dread, and others (for example, media scholars James W. Carey and John J. Quirk in their companion essays "The Mythos of the Electronic Revolution."[15,16]) instead emphasizing the spiritual or millennial hopes that come to be invested in new technologies.

† Though the aim is to promote drug development rather than serve the medical needs of prospective subjects, the policy seems calculated to increase the volume of – and hence access to – phase 1 trials.

Center, publish materials for prospective volunteers on the web that expressly describe phase 1 trials as therapeutic.[26]

An analysis of benefits and translational trials

Do patients who enter translational trials actually benefit from participation? Do researchers and policy-makers err in conceiving such trials as therapeutic? And to what degree do medical benefits ethically justify risks in first-in-human trials?

Answering these questions necessitates a careful look at what, precisely, is meant by "benefit" in clinical research. In recent years, ethicists have taken to dividing trial benefits into three categories.[32] *Aspirational benefit* refers to what has otherwise been called scientific or medical value; these accrue to persons other than the research subject. However, one might also create a category of aspirational benefits that accrue to the subject: the degree to which persons who enter clinical trials receive satisfaction from the knowledge that their participation is advancing moral ends consistent with their own.[‡] Survey data indicate that, though this objective is seldom primary in decisions to enroll in phase 1 studies, it is nevertheless frequently identified as a consideration.[33]

The second category is *collateral benefit*. This refers to unintended, though welcome, benefits received from participation in clinical trials. Collateral benefits might be further divided into medical as well as non-medical benefits, with the former comprising improved monitoring and symptom control, and the latter involving psychological comfort deriving from the preservation of terminal patients' hope. The medical literature provides some evidence of both types of benefit. For example, several studies indicate that subjects, on average, have better outcomes than persons treated outside of research protocols.[34,35] Evidence supporting such a "trial effect" is not, however, particularly strong.[36] One limitation of studies showing trial effects is that observed medical gains from study participation might reflect a tendency for more resilient, invested, or better-off patients to enroll in trials. Moreover, to my knowledge, there are no analyses indicating that phase 1 study patients in particular derive measurable collateral medical benefits compared with otherwise eligible patients who decline participation. With respect to effects on morale, various data gathered in the phase 1 oncology setting show important psychological benefits, including a sense of taking control,[33] sustained hope,[37] comfort in maintaining a therapeutic alliance,[38] and diminished anxiety.[39] These advantages are somewhat offset by nonspecific burdens of trial participation, including logistical demands, schedule disruption, transportation to a trial site,[40] and the emotional turmoil when a study ends.[41]

A third way that translational trials benefit subjects is through *direct benefit* – that is, by response to the study drug. Certainly, patients have experienced major clinical improvement during translational trials.[§] Some commentators cite initial studies of cisplatin, in which as many as 25 percent of patients were cured of testicular cancer.[42,43] More recently, 64 percent of patients enrolled in a phase 1 dose-escalating study of imatinib mesylate (known by the

---

[‡] Nancy King, who formulated these three categories of benefits, would classify such "gratification" as a collateral benefit. I think there are good reasons to consider it separately. As I argue below, collateral benefits are ethically suspect as inducements to trial participation. Moral gratification, however, seems like the most defensible form of inducement – provided, of course, that such gratification is grounded on a credible claim of medical value.

[§] Note that all data discussed below derived from the oncology setting, and thus might not reflect other types of phase 1 trials. Within GT, for example, phase 1 studies involving X1-SCID, ADA-SCID, and CGD are examples of studies that produced cures.

tradename Gleevec®) showed complete responses.[44] Two large meta-analyses have been performed recently on phase 1 oncology studies. One found rates of objective tumor response – that is, major shrinkage of a tumor – of 3.8 percent (3.5 percent for biologics and targeted agents, which are more typical of translational studies) and trial death rates of 0.54 percent (0.19 percent for targeted and biologicals).[45] The other similarly found response rates of 4.8 percent, and death rates of 0.26 percent for first-in-human studies.[46] This latter study also reported a small sample of gene transfer studies: 0 percent response, 0 percent death.

Evaluating appeals to aspirational and collateral benefit

On the whole, the claim that phase 1 study volunteers benefit from the experience – whether from fulfillment of aspirations, collaterally, or directly – and that these benefits generally outweigh the risks of trial participation is a credible one. It might be tempting, then, to conclude that investigators and IRBs can put aside categorical concerns about risk in translational trials on the view that these are partly or entirely compensated for by their aspirational, collateral, and/or direct medical benefits. In the next several sections, I want to show why this reasoning is flawed.

The position that study risks are compensated by aspirational benefits is uncontroversial. By definition, research is justified insofar as it serves the goal of producing generalizable knowledge. In an influential essay published four decades ago, philosopher Hans Jonas argued that human experiments are the most ethically defensible where the moral objectives of clinical investigators align with those of volunteers.[❡] The benefits to society, as well as the fulfillment of a volunteer's moral yearnings, are thus paramount considerations in justifying risk.

Using collateral benefits to justify a study's risks and burden is more problematic.[48] Surveys show that many patients and investigators in early-phase studies believe that trials provide meaningful collateral benefits through psychosocial support and personal relationships with study teams.[49] That volunteers enter trials seeking such benefits (and that investigators deliberately provide them) is not in itself cause for ethical alarm (see section "Therapeutic ambiguity, patient motivations, and informed consent", below). The question here is whether such benefits can be said to ethically justify risks and burdens associated with trials. I think there are several reasons to answer no. First, doing so would create room for a lot of mischief by providing a mechanism whereby investigators and IRBs can rationalize procedures that are primarily intended for research purposes without having to defend their scientific value. For example, positron emission tomography (PET) to track the dissemination of a vector in a patient's brain offers a small, but non-zero, possibility of detecting brain tumors that are amenable to treatment.[50,51] Yet PET scanning also exposes volunteers to modest levels of ionizing radiation, approximately equivalent to three years of background radiation and well within occupational limits in the nuclear industry. Whether the benefits of detecting a rare, treatable brain tumor outweigh the radiological risks is ambiguous and debatable. On a purely pragmatic level in terms of providing investigators and IRBs with guidance, it would be far easier to ask whether PET scans depart from standard care than to ask whether this increase in exposure to radiation presents net risks. If the procedure is a departure from standard care, the investigator should be able to justify it by describing its scientific merits.

The second problem with the use of collateral benefits to offset research risk is that it is in tension with ethical medical practice. Many commentators worry about the fairness of

❡ This reasoning is consistent with practice in other realms where clinicians perform procedures that are demonstrably contrary to the direct medical interests of their patients but are nevertheless consistent with their moral ones. Examples include the use of healthy organ donors and circumcision of Jewish boys.[47]

embedding research in suboptimal clinical practice. That is, volunteers should not be forced to enter research studies in order to receive appropriate medical monitoring.[32] Advantaging persons by providing extra care conditional on trial participation means, in effect, disadvantaging those who decline participation.

A third objection concerns a particular type of collateral benefit that is frequently invoked to justify risk in translational trials: the psychological benefits of enrollment, including the preservation of a dying patient's hope. This well-intended way of justifying risks conflicts with how medicine and society assess the risk of therapeutic interventions. For example, clinicians generally justify potent drugs or surgeries not because they "preserve hope," but rather because specific, physiologically mediated therapeutic effects are postulated. Contemporary medical practice doesn't even sanction benign practices like therapeutic placebos, the laying on of hands, or homeopathic potions. Thus, while encouraging hope might be a benefit for participation in a phase 1 clinical trial, this should not enter the ethical evaluation of a study's risks.

Evaluating arguments for translational trials as therapy

Far more difficult is the question of whether direct benefits partly or completely justify the risks of participating in translational trials. As sketched in Chapter 3, the standard approach to the ethical analysis of risk in clinical research is to divide study interventions into those that have therapeutic warrant (like study drugs in a randomized controlled trial) and those that do not (e.g. a series of venipunctures to monitor the pharmacokinetics of a study drug). Risks of the former are evaluated on the basis of equipoise: if administration of any study agent in a protocol is inconsistent with expert opinion on appropriate medical care, the study is unethical. Risks of procedures that lack therapeutic warrant are weighed against the social value of the anticipated gain in knowledge.**

In most clinical trials, the demarcation of research procedures into "therapeutic" and "non-therapeutic" components is rather straightforward. However, neither the originators of component analysis,[52] those who have elaborated and adapted it,[48,53,54] nor those who have put forward a modified approach[55] have ever specified how component analysis might be applied to phase 1 trials involving patients.

One approach would view phase 1 trials as either consistent with competent medical care or offering net benefits for patients. On this view, the ethical assessment of risk in a gene-transfer study will depend on a determination of whether, on balance, the risks of receiving a study agent are outweighed by its potential for direct benefit. Many oncologists,[29,30,56–59] and some ethicists[42,60,61] seem to either implicitly or explicitly endorse this position. Proponents of this view offer several defenses.

The argument from data

Data cited above (see section "An analysis of benefits and translational trials") might justify the claim that phase 1 studies have therapeutic value. Examples like imanitib mesylate

** Note that, in the previous section, I classified screening biopsies as a procedure that lacked therapeutic warrant. A critical reader might object that these biopsies are not performed to produce knowledge, but rather, to assure safety; as such, it makes no sense to weigh the biopsy risk against the value of knowledge gained. However, this objection seems to miss the point. The value of biopsies lies not in their ability to directly produce generalizable knowledge, but instead in their ability to enable the production of generalizable knowledge. It is thus acceptable to consider that value comprises both direct and indirect components.

provide further support. But it is worth considering the methodological assumptions and limitations of the studies that measure the risks and direct benefits of phase 1 cancer studies. For example, one meta-analysis involved only published phase 1 studies. Over a third of such studies go unpublished,[62,63] which means that the sample from which estimates were derived is likely biased in favor of benefits. The other meta-analysis included trials from the National Cancer Institute Cancer Therapy Evaluation Program (CTEP). Though this avoids the non-publication problem (since the authors were able to access unpublished data) it introduces a new one, namely, that the sample could represent an elite cross-section of phase 1 studies. Additionally, both studies measure benefit using surrogate endpoints (that is, tumor shrinkage rather than actual increases in survival). Whether such physiological measures translate into the kinds of clinical improvement that prospective subjects desire – prolongation of life or diminished symptoms – is unclear.[64–70]

Using instances like imatinib meslyate phase 1 trials to support a claim that all phase 1 trials have therapeutic value is problematic. It seems contradictory to make a general rule – that phase 1 trials are therapeutic – on the basis of exceptional cases.

The argument from design

Contrary to what is often argued, Phase 1 trials frequently embody therapeutic objectives. No one would deny that their primary intention is the collection of information about a drug's safety, pharmacokinetics, and, increasingly, efficacy through the testing of biomarkers. Nevertheless, such studies often incorporate design features intended to provide enrolling subjects with direct benefits.

Various innovative dose-escalation regimes illustrate this point. Classic phase 1 studies typically enroll three to six patients in each cohort; as a result, many subjects entering studies actually receive sub-therapeutic doses. Within oncology, however, several alternative dosing strategies have been proposed that increase the probability that volunteers will receive biologically active doses. Accelerated titration designs, for example, sometimes enroll one subject in each cohort until toxic effects are observed, or increase dosing levels rapidly.[71] They also sometimes allow for patients in early cohorts to receive larger drug doses later in the study (intra-patient dose escalation). Another approach, the continual reassessment method, modifies dosing levels as a study proceeds using statistically derived projections from the response of initial subjects.[72] A related set of study reforms include the use of biomarkers or other patient characteristics to customize dose levels administered in a phase 1 study.[73] One final reform, cohort-specific consent, allows research subjects to select their study dose on the basis of information about previous subjects (in one study, the first subject was given the option of three different doses).[74,75]

The argument from design, however, runs into a series of conceptual and empirical problems. With respect to the former, phase 1 trials embody design features that militate against therapeutic activity for all volunteers. This is because typical designs aim at defining an envelope of therapeutic activity, and this requires that some patients receive doses below therapeutic levels (sub-therapeutic doses) and others at toxic levels ("dose limiting toxicity").[††]

[††] There are exceptions: in an era of targeted therapeutics, trialists are abandoning endpoints based on the maximum tolerated dose in favor of maximum effective dose. And some translational trials in gene transfer (for example, studies involving SCID and Parkinson's disease) avoid the use of toxicity endpoints. Nevertheless, the familiar (if grandfatherly) admonition "somebody's going to get hurt" often holds for many other translational trials.

The fact that patients in middle cohorts receive active and tolerable doses does not change or, in itself ethically redeem the fact that other patients, by virtue of study structure, receive an investigational drug in a manner that is not consistent with therapeutic action.[76]

The empirical problem with the argument from design is that preliminary data suggest that reforms intended to improve the therapeutic value of phase 1 studies have been counterproductive. A recent meta-analysis of phase 1 cancer studies using novel dose-escalation regimes showed significantly more hematological adverse events than standard dosing regimes, without corresponding improvements in disease response. It also found no significant differences in risks or benefits for studies involving intra-patient dose escalation.[77‡‡] Another recent meta-analysis reinforced these findings.[78] An experiment with cohort-specific consent also produced infelicitous results: study subjects piled into a specific cohort after one patient at the dosing level experienced a complete response. Had a conventional phase 1 dose-escalation scheme been used, fewer subjects would have been necessary to complete the study, and probably fewer toxicities would have been observed.[75§§]

The argument from equipoise

Clinical equipoise requires that volunteers do not receive a treatment that is known by the expert community to be inferior to standard of care.[79] That many expert oncologists regard phase 1 studies as a therapeutic activity would suggest de facto evidence of clinical equipoise. Moreover, equipoise requires that a study intervention be as clinically useful as the standard of care. For the treatment-refractory patients that are typically enrolled in phase 1 gene-transfer studies, the alternative to enrollment is no further treatment. Under this conception of equipoise, risks in translational trials are morally underwritten by a potential for direct benefit.

A more careful reading of the concept of clinical equipoise, however, shows this reasoning to be flawed.[76] In Freedman's original formulation of clinical equipoise, he speaks of "the reason for conducting clinical trials: there is a current or immanent conflict in the clinical community ..." over the therapeutic value of an intervention.[79] If "current or immanent" conflict grounds a claim of equipoise at the point of randomized controlled trial initiation, how can it be said to exist at two steps earlier in the process, where evidence levels grounding a claim of clinical equipoise are weaker? Proponents of equipoise argue that clinical decision-making requires a "robust epistemic threshold."[80] That is, sound medical practice adopts a skeptical and guarded position towards guesswork and intuition, basing medical decision-making instead on reliable evidence. The high attrition rate of drugs entering phase 1 trials would seem to suggest that investigators are very bad

‡‡ There are several important limitations of this study. First, the authors only examined cytotoxic drugs. Second, direct benefit was determined on the basis of response rates rather than clinical endpoints. Third, the meta-analysis was based on published studies. As a result, publication bias might distort their findings.

§§ I nevertheless believe that such dosing strategies are not ethically inappropriate, because they reconcile tensions between research and therapy in phase 1 studies. That is, standard dosing regimes accentuate the divergence of motivations between volunteers, who enter phase 1 studies with therapeutic objectives, and investigators, who pursue such studies primarily seeking generalizable knowledge. Hans Jonas long ago argued that human experimentation is least worrisome where the objectives of clinical investigators and volunteers align.[47]

guessers when it comes to efficacy for novel interventions¶. Medical ethics is founded on the Hippocractic dictum, "at least, do no harm." The default position consistent with this ethic would be to assume that new agents are ineffective and possibly even harmful unless there is clinical evidence to the contrary.

The argument from consistency with policy

Finally, various policies seem to tacitly endorse the conception of phase 1 trials as a venue for curative therapy. US Medicare, for example, denies coverage of hospice care for patients entering phase 1 studies, because such trials are viewed as aimed at cure.[82] In addition, pediatric phase 1 oncology studies are generally classified as research "presenting the prospect of direct benefit to the individual subjects" under subpart D of the US Common Rule.[60,83]

But policy is divided. For example, the notion that phase 1 trials have therapeutic justification is inconsistent with prevailing therapeutic standards. It is nearly inconceivable that the FDA would ever approve a drug that had the properties of phase 1 cancer studies (e.g. methodologically flawed preclinical evidence of biological activity – see Chapters 6 and 7) or their rates of clinical success (e.g. no demonstration of clinical evidence of efficacy, non-validated surrogates data indicating that one in twenty patients show a tumor response, and a rate of severe toxicity in the 10 percent range). Moreover, therapeutic benefit is not a major concern when the FDA reviews applications to conduct phase 1 studies of new drugs. True, sponsors are required to submit information on the pharmacology and predicted mechanisms of action of the new drug. But in the words of the FDA guidance clarifying this requirement, "The regulations do not further describe the presentation of these data [on pharmacology and drug distribution], in contrast to the more detailed description of how to submit toxicologic data ... To the extent that such studies may be important to address safety issues, or to assist in evaluation of toxicology data, they may be necessary; however, lack of this potential effectiveness information should not generally be a reason for a Phase 1 IND [Investigational New Drug] to be placed on clinical hold."[84]

Another area where policy would seem not to favor the view of phase 1 studies as therapy is in health insurance: many private and public programs in the USA routinely cover clinical trial costs, but explicitly exclude phase 1 cancer studies unless the agent is approved for a different indication.[85,86]

Translational trials and the justification of risk

Many commentators express views that are more consistent with the position that administration of study drugs in phase 1 trials should be viewed as a "research procedure" such that risks are evaluated entirely on the basis of benefits to future patients.[87–91]

I nevertheless favor a more nuanced position that accommodates the heterogeneity of phase 1 trial design. The phase 1 trial category includes first-in-human studies, refinement studies (that is, testing a new agent against a particular type of cancer, or testing a reformulated version of the new agent), combination studies involving one unapproved drug, studies combining two approved drugs, and pediatric studies. Moreover, within the first-in-human study category, many involve agents that are similar to those that have already been tested, and only stereotyped translational trials involve radically new interventions. I favor an approach that classifies phase 1 study interventions according to whether they are novel or not. Those in the former category are classified as therapeutically unwarranted in risk assessment, such that

¶ According to arecent study, 27 percent of cancer drugs entering phase 1 studies are eventually licensed. I suspect the number is much smaller for novel agents[81].

risks are justified to the extent that the study serves social ends. This would furnish innovative translational trials with a powerful incentive to maximize scientific value.

On what basis would interventions be designated as novel? Minimal criteria that should be considered include: whether the intervention involves one of the first attempts to target a particular biological pathway (e.g. first attempt to activate a certain T-cell receptor), one of the first uses of a new vector (e.g. first use of a measles vector), one of the first uses of a new transgene (i.e. a transgene that is not significantly homologous to others tested in other trials), one of the first attempts to target a particular tissue (e.g. the first delivery of vector to the subthalamic nucleus), or one of the first attempts to use a new platform against a disease (e.g. the first gene-transfer trial against multiple sclerosis).

Non-novel interventions would be evaluated on a case-by-case basis as to whether they are compatible with therapeutic warrant. The default position here should be that, for the purposes of risk assessment, these studies are grounded in research warrant rather than therapeutic warrant. This proposition is based on arguments from design and equipoise articulated in the previous section. However, if evidence and clinical practice support therapeutic expectation for all subjects, a claim of therapeutic warrant might be justified. This would occur in studies where the clinical equipoise condition of immanent uncertainty is met, and where design involves escalation from doses that are widely understood to have therapeutic activity. In such circumstances, studies become subject to a more permissive value standard.

Therapeutic ambiguity, patient motivations, and informed consent

In the previous section I argued that various institutions arrayed around translational trials should exercise caution in interpreting translational phase 1 trials as therapeutic endeavors. But to what extent should we worry when patients interpret them as such?

Numerous studies show that patients enter phase 1 trials for care. That they should so is not surprising. I previously noted that investigators often view phase 1 trial participation as a therapeutic option. Within gene transfer, one study found that only 43 percent of principal investigators in phase 1 trials did not expect that volunteers would benefit directly.[92] More troubling, patients often estimate the likelihood of clinical benefits as significantly higher than the clinicians who enroll them.[23,93,94] These observations raise important questions about the validity of informed consent: if volunteers enroll with unrealistic expectations about clinical benefits (therapeutic overestimation),[95] or without understanding that trials are structured in ways that can interfere with therapeutic objectives (the therapeutic misconception),[96,97] might they make decisions that are inconsistent with their goals?

Some commentators take a “contrarian”[98] view, arguing that therapeutic misconception does not raise major concerns where trial participation is not expected to have major consequences for care.[95,99,100] But those who make this argument tend to disavow the claim for early-phase trials, where the risk–benefit balance is less favorable for volunteers and departure from standard of care is more significant. I therefore take it as axiomatic that, because therapeutic expectations can lead patients to enroll in studies that might run contrary to their considered preferences, florid therapeutic misconception and overestimation in early-phase studies point to important concerns about consent validity.

Kevin Weinfurt and colleagues raise other questions about the ethical significance of therapeutic expectations in early-phase cancer studies. They note that survey questions about therapeutic expectations elicit at least two possible categories of statements: those about frequency and those about belief. Frequency statements concern how likely a patient believes a population will benefit from an intervention. Belief statements concern the likelihood that, as individuals, subjects expect to receive benefits. Whereas frequency statements are falsifiable, belief statements are not, and volunteers have many reasons for confidently asserting benefit expectation, such as a desire to project optimism.[101–103] There are thus sound reasons to question the ethical significance of therapeutic misconception and overestimation.[¶]

Nevertheless, in one study, Weinfurt and colleagues put a question to volunteers that forced them to provide frequency statements. Phase 1 trial subjects, on average, stated that they expected that 50 of 100 patients would "have their cancer controlled as a result of the experimental therapy."[103] These results do not exclude the possibility that volunteers provide such optimistic assessments out of "belief" rather than "probabilistic expectation;" they nevertheless keep alive the possibility that persons enrolled in early-phase trials overestimate the probability of benefit.

Another set of commentators have raised questions about the moral significance of therapeutic expectations by making a deterministic argument. According to this view, therapeutic misunderstandings are inevitable because of cognitive processes that are inaccessible to investigators conducting informed-consent trials.[105,106] Researchers might dampen therapeutic expectations by manipulating the affect of volunteers. One common proposal is to offer small sums of money. Another is to ask investigators to wear special uniforms. A third, slightly amusing proposal is to have "unattractive staff" solicit consent. Ultimately, however, proponents of these proposals despair that misunderstandings cannot be eradicated.

Finally, interviews with researchers and subjects in gene-transfer trials reveal some of the intricacy and subjectivity in demarcating "care" and "research." Research subjects frequently recognize that studies are primarily intended to serve medical advance, yet view investigators as caregivers. When asked to explain why they perceived studies as providing care, volunteers often answered that they believed that investigators provided both physical and emotional support. Similarly, about half of investigators saw their care-giving relationship with volunteers to be of equal or greater significance than their researcher functions.[49] These findings suggest that, in many circumstances, volunteers have highly nuanced understandings of the intentions of researchers and the implications of their participation in studies. Nevertheless, a related study involving similar patients showed that they frequently entered studies with strong expectations of direct benefit, and/or believing that the study is primarily intended to help them.[25]

Clearly, further research is needed to better understand the basis, malleability, meaning, and ethical implications of therapeutic expectation in early-phase research. Until then, some basic prescriptions can be followed. Just as ethicists should not uncritically take the position that all therapeutic expectations observed in surveys are morally questionable, nor should ethicists and others assume that treatment-refractory patients are able to "think clearly" and "choose freely,"[42] and that we should unreservedly respect their therapeutic interpretations. A precondition for the exercise of autonomy is the existence of a stable set of values and preferences. In many instances, a volunteer's therapeutic position might reflect longstanding values like optimism, perseverance, and faith in medical science, in which case such preferences ought to be respected.[100] In other circumstances, however, the preferences of the critically ill subject might not meet this precondition. Health-related values and

[¶] Another factor that further muddies this question is the behavioral economics finding that persons facing significant expected losses (in this case, death from cancer) assign different value to remote outcomes than do persons not confronting significant expected losses. Responses in surveys could be measuring a legitimate shift in values rather than a misunderstanding.[104]

Table 5.1 Recommendations for assessment and disclosure of benefit in translational trials

Ethical review of risks in phase 1 trials
For novel interventions, risks should be ethically justified by appeals to social value
For non-novel interventions, risks can be justified by appeals to social value or direct benefits, depending on the case
Risks should not be justified by appeals to collateral benefits
Informing volunteers
Unlikely probability of major clinical impact should be disclosed
Describe benefits as clinical endpoints, or else explain uncertain relationship between surrogate and clinical endpoints
Procedures that interfere with therapeutic objectives (e.g. subtherapeutic dosing) should be disclosed
Investigators should explore the motivations of volunteers who enter phase 1 studies to dispel misunderstandings

preferences vary significantly according to the individual's medical status. It follows that a person's values and preferences will undergo significant changes as he or she descends from health, through shock, into acceptance.*** This instability should at least register concern for anyone committed to a rich conception of autonomy.

Investigators and ethics committees can implement measures to thwart therapeutic misestimations and misconceptions (Table 5.1). The goal here is not necessarily to disabuse subjects of their therapeutic motivations or therapeutic beliefs, but rather to ensure that they appreciate the information they need to decide whether a given trial will adequately advance their objectives.

First, volunteers should be informed that major clinical benefits of trial participation are highly improbable – that most agents that enter clinical testing are abandoned because they prove unsafe or ineffective. This recommendation is consistent with what the NIH already recommend for gene-transfer studies.††† Currently, less than a third of consent documents in phase 1 gene-transfer trials declare that any direct benefit – for example, cure or major clinical improvement – is improbable.[92,108] Instead, consent documents tend to offer vague descriptions, like the statement "we cannot guarantee this drug will shrink your tumors." Statements like this contain virtually no information about the probability or magnitude of benefits, and are logically compatible with a view that 99.9 percent of volunteers will benefit. In contrast, the statement "participating in this study is unlikely to result in major clinical improvement" is far more informative and accurately conveys current knowledge from the appropriate reference class about the benefits of translational trials. Along a similar vein, consent forms tend toward therapeutically loaded language by using terms like "treatment" and "gene therapy" instead of "study agent" or "gene transfer."[109]

*** According to prospect theory, decision-makers value outcomes relative to a reference point. As noted by Kevin Weinfurt, patients newly diagnosed with a terminal illness might typically be using their pre-diagnosis reference point when making value assessments for a given treatment regime. As more time passes, patients will typically settle on a post-diagnosis reference point that values small or improbable survival gains from trial participation differently.[104] The point I'm making is that, where phase 1 trials typically involve newly diagnosed patients (e.g. cancer), value regimes for volunteers will often be in flux.

††† "Consent document should clearly state that no direct clinical benefit to subjects is expected to occur as a result of participation in the study, although knowledge may be gained that may benefit others."[107]

Second, investigators should distinguish between clinical benefits and surrogate benefits when describing protocols to volunteers. One study of gene-transfer consent practices found that investigators typically described benefits to volunteers in terms of surrogates like tumor shrinkage (for cancer) or blood vessel formation (for studies involving cardiovascular disease).[92] When possible surrogate benefits are described, investigators should explain the uncertain relationship between surrogate outcomes and meaningful clinical benefit.

Third, all subjects should be informed of any procedures or practices that will interfere with their therapeutic objectives. For example, investigators should always inform subjects that there is a possibility of receiving a subtherapeutic dose, and once they are enrolled, whether they have been assigned to a subtherapeutic dose cohort; volunteers should also be told of what will happen after the trial is completed. Will they be eligible to continue receiving the study drug? Will the investigator continue to follow them? These might seem obvious prescriptions, but current informed-consent practice in gene transfer often does not describe procedures that frustrate therapeutic objectives.[110] Though consent documents used in controlled clinical trials routinely disclose the use of randomization and treatment concealment, in my experience with phase 1 consent documents, disclosure of subtherapeutic dosing or ineligibility for switching dose cohorts is uncommon.

Fourth, persons conducting informed-consent trials should ask subjects why they decided to enter the study, and if their motivations are therapeutic, explain aspects of the study that would frustrate such objectives. The investigator should further ask subjects what they think are the odds that other persons in the study will experience significant clinical improvement or major toxicities. The investigator might also query the subject about how long he or she thinks it will take before the study agent tested in the protocol becomes a standard therapy. Answers to any of these questions that are inconsistent with the clinician's (or IRB's) assessments might signal the need for further discussions with the volunteer.

Conclusions

In the end, the RAC rejected the single-patient compassionate-use protocol with which this chapter opened. But Royston and colleagues prevailed with the Department of Human Services, and were granted a "compassionate plea exemption."[111] After receiving the intervention, the woman succumbed to her illness;[112] according to the published report, her outcome "supported ... evaluation ... in additional patients."[113] Thirteen years later, Ark Therapeutics – a UK-based biotechnology company – has reportedly filed for marketing authorization with a gene-transfer product for glioblastoma.[114,115] Their intervention, however, is unrelated to the one employed by Royston.

The analysis given in this chapter provides two conclusions about this episode, and about other instances where a claim of direct or collateral patient benefit is mobilized to support translational trials. With respect to how investigators and REBs should assess risk and possible benefits, I argued that risks in first-in-human studies are justified by knowledge value rather than therapeutic benefit. In making this argument, however, I conceded that there may be circumstances where phase 1 studies have therapeutic warrant. However, the San Diego compassionate-use study would not qualify, because the intervention was highly novel.

With respect to consent, I offered a mixed appraisal of whether patients are justified in seeking medical care through phase 1 studies. There is nothing about phase 1 studies that categorically excludes the possibility of direct medical benefit. Nor is there necessarily a problem if volunteers enter studies seeking collateral benefits. There is no way of knowing how well Royston's patient (and her husband) understood the study's purpose and risks, and whether their therapeutic expectations were based on justified beliefs, or correct frequency estimates. I nevertheless believe that the harms to volunteers entering studies with inappropriate therapeutic expectations exceed the burdens associated with trying to prevent misunderstandings. Translational researchers should strive to provide accurate information about the improbability of significant medical benefit, and they should explain how research procedures can logically frustrate a volunteer's therapeutic objectives.

For the last several decades, medicine and ethics have expressed discomfort with the practice of performing studies that offer very poor chances for fulfilling the needs of dying patients, and favorable odds for serving the needs of future patients, investigators, and sponsors. Yet medicine cannot progress unless we continue to do so. Should the day ever arrive when research on the treatment-refractory ceases to give us pause, we'll know we have a problem.

References

1. Stupp R, Mason WP, van den Bent MJ, Weller M, Fisher B, Taphoorn MJ, *et al.* Radiotherapy plus concomitant and adjuvant temozolomide for glioblastoma. *N Engl J Med* 2005; **352**(10): 987–96.
2. Recombinant DNA Advisory Committee (RAC). Minutes of Meeting December 4, 1992; 78.
3. Recombinant DNA Advisory Committee (RAC). Minutes of Meeting December 4, 1992; 76.
4. Recombinant DNA Advisory Committee (RAC). Minutes of Meeting December 4, 1992.
5. Churchill LR, Collins ML, King NM, Pemberton SG, Wailoo KA. Genetic research as therapy: implications of "gene therapy" for informed consent. *J Law Med Ethics* 1998; **26**(1): 3, 38–47.
6. Childress JF. Nuremberg's legacy: some ethical reflections. *Perspect Biol Med* 2000; **43**(3): 347–61.
7. Epstein S. Impure science: AIDS, activism, and the politics of knowledge. Berkeley: University of California Press, 1996.
8. Edgar H, Rothman DJ. New rules for new drugs: the challenge of AIDS to the regulatory process. *Milbank Q* 1990; **68** (Suppl. 1): 111–42.
9. Cohen PD, Herman L, Jedlinski S, Willocks P, Wittekind P. Ethical issues in clinical neuroscience research: a patient's perspective. *Neurotherapeutics* 2007; **4**(3): 537–44.
10. Rabeharisoa V. The struggle against neuromuscular diseases in France and the emergence of the "partnership model" of patient organisation. *Soc Sci Med* 2003; **57** (11): 2127–36.
11. Epstein S. Democracy, Expertise, and AIDS Treatment Activism. In: Kleinman DL, ed. *Science, Technology, & Democracy*. Albany, State University of New York Press, 2000; 15–32.
12. Kasson JF. *Civilizing the Machine: Technology and Republican Values in America, 1776–1900*. New York, NY, Penguin Books, 1977.
13. Nye DE. *American Technological Sublime*. Cambridge, MA, MIT Press, 1996.
14. Miller P. *The Life of the Mind in America: From the Revolution to the Civil War*. New York, Harcourt, Brace and the World, 1965; 287–308.
15. Carey JW, Quirk JJ. The Mythos of the Electronic Revolution. Part 1. *American Scholar* 1970; **39**(2): 219–41.
16. Carey JW, Quirk JJ. The Mythos of the Electronic Revolution. Part 2. *American Scholar* 1970; **39**(3): 395–424.
17. Marx J. Cancer drug's fickle nature explained. *Science Now* (Online) 2004, April 29.
18. US Food, and Drug Administration. *Drugs for Human Use: Investigational New Drug Application 21CFR312.34*. US Department of Health and Human Services, 2007.
19. US Food, and Drug Administration. *Guidance for Industry: INDS – Approaches to Complying with CGMP During Phase 1* (Draft Guidance). US Department of Health and Human Services, 2006.
20. National Institutes of Health. Fact Sheet: ClinicalTrials.gov. Bethesda, MD, National Library of Medicine, 2005.

21. Talarico L, Pazdur R. Expanded access program (EAP) to investigational drugs. *J Clin Oncol (*Meeting Abstracts*)* 2004; **22**(14_suppl.): 6001.
22. Daugherty CK, Banik DM, Janish L, Ratain MJ. Quantitative analysis of ethical issues in phase I trials: a survey interview of 144 advanced cancer patients. *IRB* 2000; **22**(3): 6–14.
23. Daugherty C, Ratain MJ, Grochowski E, Stocking C, Kodish E, Mick R, *et al.* Perceptions of cancer patients and their physicians involved in phase I trials. *J Clin Oncol* 1995; **13**(5): 1062–72.
24. Nurgat ZA, Craig W, Campbell NC, Bissett JD, Cassidy J, Nicolson MC. Patient motivations surrounding participation in phase I and phase II clinical trials of cancer chemotherapy. *Br J Cancer* 2005; **92**(6): 1001–5.
25. Henderson GE, Easter MM, Zimmer C, King NM, Davis AM, Rothschild BB, *et al.* Therapeutic misconception in early phase gene transfer trials. *Soc Sci Med* 2006; **62**(1): 239–53.
26. Hutchison C. Phase I trials in cancer patients: participants' perceptions. *Eur J Cancer Care (Engl)* 1998; **7**(1): 15–22.
27. Winerip M. Fighting for Jacob. *The New York Times* 1998 December 6.
28. Eisenhauer E, Twelves C, Buyse M. *Phase 1 Cancer Clinical Trials: A Practical Guide.* New York, NY, Oxford University Press, 2006.
29. Markman M. "Therapeutic intent" in phase 1 oncology trials: a justifiable objective. *Arch Intern Med* 2006; **166**(14): 1446–8.
30. Khandekar J, Khandekar M. Phase 1 clinical trials: not just for safety anymore? *Arch Intern Med* 2006; **166**(14): 1440–1.
31. Joffe S, Weeks JC. Views of American oncologists about the purposes of clinical trials. *J Natl Cancer Inst* 2002; **94**(24): 1847–53.
32. King NM. Defining and describing benefit appropriately in clinical trials. *J Law Med Ethics* 2000; **28**(4): 332–43.
33. Moore S. A need to try everything: patient participation in phase I trials. *J Adv Nurs* 2001; **33**(6): 738–47.
34. Braunholtz DA, Edwards SJ, Lilford RJ. Are randomized clinical trials good for us (in the short term)? Evidence for a "trial effect". *J Clin Epidemiol* 2001; **54**(3): 217–24.
35. Rajappa S, Gundeti S, Uppalapati S, Jiwatani S, Abhyankar A, Pal C, *et al.* Is there a positive effect of participation on a clinical trial for patients with advanced non-small cell lung cancer? *Indian J Cancer* 2008; **45**(4): 158–63.
36. Peppercorn JM, Weeks JC, Cook EF, Joffe S. Comparison of outcomes in cancer patients treated within and outside clinical trials: conceptual framework and structured review. *Lancet* 2004; **363**(9405): 263–70.
37. Barrera M, D'Agostino N, Gammon J, Spencer L, Baruchel S. Health-related quality of life and enrollment in phase 1 trials in children with incurable cancer. *Palliat Support Care* 2005; **3**(3): 191–6.
38. Cox K. Assessing the quality of life of patients in phase I and II anti-cancer drug trials: interviews versus questionnaires. *Soc Sci Med* 2003; **56**(5): 921–34.
39. Berdel WE, Knopf H, Fromm M, Schick HD, Busch R, Fink U, *et al.* Influence of phase I early clinical trials on the quality of life of cancer patients. A pilot study. *Anticancer Res* 1988; **8**(3): 313–21.
40. Cohen MZ, Slomka J, Pentz RD, Flamm AL, Gold D, Herbst RS, *et al.* Phase I participants' views of quality of life and trial participation burdens. *Support Care Cancer* 2007; **15**(7): 885–90.
41. Cox K. Researching research: Patients' experiences of participation in phase I and II anti-cancer drug trials. *Eur J Oncol Nurs* 1999; **3**(3): 143–52.
42. Agrawal M, Emanuel EJ. Ethics of phase 1 oncology studies: reexamining the arguments and data. *JAMA* 2003; **290**(8): 1075–82.
43. Higby DJ, Wallace HJ Jr, Albert DJ, Holland JF. Diaminodichloroplatinum: a phase I study showing responses in testicular and other tumors. *Cancer* 1974; **33**(5): 1219–25.
44. Druker BJ, Talpaz M, Resta DJ, Peng B, Buchdunger E, Ford JM, *et al.* Efficacy and safety of a specific inhibitor of the BCR-ABL tyrosine kinase in chronic myeloid leukemia. *N Engl J Med* 2001; **344**(14): 1031–7.
45. Roberts TG Jr, Goulart BH, Squitieri L, Stallings SC, Halpern EF, Chabner BA, *et al.* Trends in the risks and benefits to patients

with cancer participating in phase 1 clinical trials. *Jama* 2004; **292**(17): 2130–40.

46. Horstmann E, McCabe MS, Grochow L, Yamamoto S, Rubinstein L, Budd T, *et al.* Risks and benefits of phase 1 oncology trials, 1991 through 2002. *N Engl J Med* 2005; **352**(9): 895–904.
47. Jonas H. Philosophical Reflections on Human Experimentation. In: Freund PA, ed. *Experimentation with Human Subjects.* London, George Allen & Unwin, 1972.
48. National Bioethics Advisory Commission. *Ethical and Policy Issues in International Research: Clinical Trials in Developing Countries.* Bethesda, MD, National Bioethics Advisory Commission, 2001.
49. Easter MM, Henderson GE, Davis AM, Churchill LR, King NM. The many meanings of care in clinical research. *Sociol Health Illn* 2006; **28**(6): 695–712.
50. The example in this paragraph is a modified version of one offered by David Wendler and Franklin Miller in Wendler D, Miller FG. Assessing research risks systematically: the net risks test. *J Med Ethics* 2007; **33**(8): 481–6.
51. Mamourian A. Incidental Findings on Research Functional MR Images: Should We Look? In: Glannon W, ed. *Defining Right and Wrong in Brain Science: Essential Readings in Neuroethics.* San Francisco, CA, Dana Press, 2002.
52. Freedman B, Fuks A, Weijer C. Demarcating research and treatment: a systematic approach for the analysis of the ethics of clinical research. *Clin Res* 1992; **40**(4): 653–60.
53. Weijer C. The ethical analysis of risk. *J Law Med Ethics* 2000; **28**(4): 344–61.
54. Weijer C, Miller PB. When are research risks reasonable in relation to anticipated benefits? *Nat Med* 2004; **10**(6): 570–3.
55. Wendler D, Miller FG. Assessing research risks systematically: the net risks test. *J Med Ethics* 2007; **33**(8): 481–6.
56. American Society of Clinical Oncology. Critical role of phase I clinical trials in cancer treatment. *J Clin Oncol* 1997; **15**(2): 853–9.
57. Lipsett MB. On the nature and ethics of phase I clinical trials of cancer chemotherapies. *JAMA* 1982; **248**(8): 941–2.
58. Muggia FM. Phase 1 clinical trials in oncology. *N Engl J Med* 2005; **352**(23): 2451–3; author reply, 2451–3.
59. Kurzrock R, Benjamin RS. Risks and benefits of phase 1 oncology trials, revisited. *N Engl J Med* 2005; **352**(9): 930–2.
60. Kodish E. Pediatric ethics and early-phase childhood cancer research: conflicted goals and the prospect of benefit. *Account Res* 2003; **10**(1): 17–25.
61. Ackerman TF. Phase 1 pediatric oncology trials. *J Pediatr Oncol Nurs* 1995; **12(3)**: 143–5.
62. Chen EX, Tannock IF. Risks and benefits of phase 1 clinical trials evaluating new anticancer agents: a case for more innovation. *JAMA* 2004; **292**(17): 2150–1.
63. Camacho LH, Bacik J, Cheung A, Spriggs DR. Presentation and subsequent publication rates of phase I oncology clinical trials. *Cancer* 2005; **104**(7): 1497–504.
64. Buyse M, Thirion P, Carlson RW, Burzykowski T, Molenberghs G, Piedbois P. Relation between tumour response to first-line chemotherapy and survival in advanced colorectal cancer: a meta-analysis. Meta-Analysis Group in Cancer. *Lancet* 2000; **356**(9227): 373–8.
65. Epstein S. Activism, drug regulation, and the politics of therapeutic evaluation in the AIDS era: a case study of ddC and the 'surrogate markers' debate. *Soc Stud Sci* 1997; **27**(5): 691–726.
66. Johnson KR, Ringland C, Stokes BJ, Anthony DM, Freemantle N, Irs A, *et al.* Response rate or time to progression as predictors of survival in trials of metastatic colorectal cancer or non-small-cell lung cancer: a meta-analysis. *Lancet Oncol* 2006; **7**(9): 741–6.
67. Pazdur R. Response rates, survival, and chemotherapy trials. *J Natl Cancer Inst* 2000; **92**(19): 1552–3.
68. Chen TT, Chute JP, Feigal E, Johnson BE, Simon R. A model to select chemotherapy regimens for phase III trials for extensive-stage small-cell lung cancer. *J Natl Cancer Inst* 2000; **92**(19): 1601–7.
69. Fleming TR, DeMets DL. Surrogate end points in clinical trials: are we being misled? *Ann Intern Med* 1996; **125**(7): 605–13.

70. Miller FG, Joffe S. Benefit in phase 1 oncology trials: therapeutic misconception or reasonable treatment option? *Clin Trials* 2008; **5**(6): 617–23.

71. Simon R, Freidlin B, Rubinstein L, Arbuck SG, Collins J, Christian MC. Accelerated titration designs for phase I clinical trials in oncology. *J Natl Cancer Inst* 1997; **89**(15): 1138–47.

72. Eisenhauer EA, O'Dwyer PJ, Christian M, Humphrey JS. Phase I clinical trial design in cancer drug development. *J Clin Oncol* 2000; **18**(3): 684–92.

73. Babb JS, Rogatko A. Patient specific dosing in a cancer phase I clinical trial. *Stat Med* 2001; **20**(14): 2079–90.

74. Freedman B. Cohort-specific consent: an honest approach to phase 1 clinical cancer studies. *IRB* 1990; **12**(1): 5–7.

75. Daugherty CK, Ratain MJ, Minami H, Banik DM, Vogelzang NJ, Stadler WM, *et al.* Study of cohort-specific consent and patient control in phase I cancer trials. *J Clin Oncol* 1998; **16**(7): 2305–12.

76. Anderson J, Kimmelman J. Extending Equipoise to Phase 1 Trials: A Bridge Too Far? In preparation, 2009

77. Koyfman SA. *The Effects of Novel Design Strategies on the Risks and Benefits of Phase 1 Oncology Trials.* New Haven, CT, Yale University, 2006.

78. Penel N, Isambert N, Leblond P, Ferte C, Duhamel A, Bonneterre J. "Classical 3 + 3 design" versus "accelerated titration designs": analysis of 270 phase 1 trials investigating anti-cancer agents. *Invest New Drugs* 2009 January 10 [Epub ahead of print].

79. Freedman B. Equipoise and the ethics of clinical research. *N Engl J Med* 1987; **317**(3): 141–5.

80. London AJ. Clinical Equipoise: Foundational Requirement or Fundamental Error? In: Steinbock B, ed. *The Oxford Handbook of Bioethics*. New York, Oxford University Press, 2007.

81. DiMasi JA, Grabowski HG. Economics of new oncology drug development. *J Clin Oncol* 2007; **25**(2): 209–16.

82. Byock I, Miles SH. Hospice benefits and phase I cancer trials. *Ann Intern Med* 2003; **138**(4): 335–7.

83. US Food and Drug Administration. Public Welfare: *Protection of Human Subjects – Additional Protections for Children Involved as Subjects in Research 45CFR46.407.* US Food, and Drug Administration, Department of Health and Human Services, 2007.

84. US Food, and Drug Administration: Center for Drug Evaluation, and Research (CDER). *Guidance for Industry: Content and Format of Investigational New Drug Applications (INDs) for Phase 1 Studies of Drugs, Including Well-Characterized, Therapeutic, Biotechnology-derived Products.* US Food, and Drug Administration, Center for Biologics Evaluation and Research (CBER), Nov 1995.

85. Gross CP, Murthy V, Li Y, Kaluzny AD, Krumholz HM. Cancer trial enrollment after state-mandated reimbursement. *J Natl Cancer Inst* 2004; **96**(14): 1063–9.

86. Aaron HJ, Gelband H. *Extending Medicare Reimbursement in Clinical Trials.* Washington, DC, The National Academies Press, 2000.

87. Miller M. Phase I cancer trials: a crucible of competing priorities. *Int Anesthesiol Clin* 2001; **39**(3): 13–33.

88. Ross L. Phase I research and the meaning of direct benefit. *J Pediatr* 2006; **149**(1 Suppl.): S20–4.

89. Kong WM. Legitimate requests and indecent proposals: matters of justice in the ethical assessment of phase I trials involving competent patients. *J Med Ethics* 2005; **31**(4): 205–8.

90. Casarett DJ, Karlawish JH, Henry MI, Hirschman KB. Must patients with advanced cancer choose between a Phase I trial and hospice? *Cancer* 2002; **95**(7): 1601–4.

91. Martin RA, Robert JS. Is risky pediatric research without prospect of direct benefit ever justified? *Am J Bioeth* 2007; **7**(3): 12–15.

92. Henderson GE, Davis AM, King NM, Easter MM, Zimmer CR, Rothschild BB, *et al.* Uncertain benefit: investigators' views and communications in early phase gene transfer trials. *Mol Ther* 2004; **10**(2): 225–31.

93. Meropol NJ, Weinfurt KP, Burnett CB, Balshem A, Benson AB 3rd, Castel L, *et al.* Perceptions of patients and physicians regarding phase I cancer clinical trials: implications for physician–patient

communication. *J Clin Oncol* 2003; **21**(13): 2589–96.

94. Cheng JD, Hitt J, Koczwara B, Schulman KA, Burnett CB, Gaskin DJ, *et al*. Impact of quality of life on patient expectations regarding phase I clinical trials. *J Clin Oncol* 2000; **18**(2): 421–8.
95. Horng S, Grady C. Misunderstanding in clinical research: distinguishing therapeutic misconception, therapeutic misestimation, and therapeutic optimism. *IRB* 2003; **25**(1): 11–16.
96. Appelbaum PS, Roth LH, Lidz CW, Benson P, Winslade W. False hopes and best data: consent to research and the therapeutic misconception. *Hastings Cent Rep* 1987; **17**(2): 20–4.
97. Lidz CW, Appelbaum PS. The therapeutic misconception: problems and solutions. *Med Care* 2002; **40**(9 Suppl.): V55–63.
98. Appelbaum PS, Lidz CW. Twenty-five years of therapeutic misconception. *Hastings Cent Rep* 2008; **38**(2): 5–6; author reply 6–7.
99. Sreenivasan G. Does informed consent to research require comprehension? *Lancet* 2003; **362**(9400): 2016–18.
100. Kimmelman J. The therapeutic misconception at 25: treatment, research, and confusion. *Hastings Cent Rep* 2007; **37**(6): 36–42.
101. Weinfurt KP, Sulmasy DP, Schulman KA, Meropol NJ. Patient expectations of benefit from phase I clinical trials: linguistic considerations in diagnosing a therapeutic misconception. *Theor Med Bioeth* 2003; **24**(4): 329–44.
102. Weinfurt KP, Depuy V, Castel LD, Sulmasy DP, Schulman KA, Meropol NJ. Understanding of an aggregate probability statement by patients who are offered participation in Phase I clinical trials. *Cancer* 2005; **103**(1): 140–7.
103. Weinfurt KP, Seils DM, Tzeng JP, Compton KL, Sulmasy DP, Astrow AB, *et al.* Expectations of benefit in early-phase clinical trials: implications for assessing the adequacy of informed consent. *Med Decis Making* 2008; **28**(4): 575–81.
104. Weinfurt KP. Value of high-cost cancer care: a behavioral science perspective. *J Clin Oncol* 2007; **25**(2): 223–7.
105. Glannon W. Phase I oncology trials: why the therapeutic misconception will not go away. *J Med Ethics* 2006; **32**(5): 252–5.
106. Charuvastra A, Marder SR. Unconscious emotional reasoning and the therapeutic misconception. *J Med Ethics* 2008; **34**(3): 193–7.
107. National Institutes of Health. *NIH Guidance on Informed Consent for Gene Transfer Research, Appendix M-III-B-1-d: Potential Benefits*. NIH, Office of Biotechnology Activities. http://oba.od.nih.gov/oba/rac/ic/appendix_m_iiI_b_1_d.html.
108. Kimmelman J, Palmour N. Therapeutic optimism in the consent forms of phase 1 gene transfer trials: an empirical analysis. *J Med Ethics* 2005; **31**(4): 209–14.
109. King NM. Rewriting the "points to consider": the ethical impact of guidance document language. *Hum Gene Ther* 1999; **10**(1): 133–9.
110. Sankar P. Communication and miscommunication in informed consent to research. *Med Anthropol Q* 2004; **18**(4): 429–46.
111. Lysaught MT. Commentary: reconstruing genetic research as research. *J Law Med Ethics* 1998; **26**(1): 48–54.
112. Jenks S. RAC splits approval on two gene therapy protocols. *J Natl Cancer Inst* 1994; **86**(1): 13–14.
113. Sobol RE, Fakhrai H, Shawler D, Gjerset R, Dorigo O, Carson C, *et al.* Interleukin-2 gene therapy in a patient with glioblastoma. *Gene Ther* 1995; **2**(2): 164–7.
114. Osborne R. Ark floats gene therapy's boat, for now. *Nat Biotechnol* 2008; **26**(10): 1057–9.
115. Cerepro® (sitimagene ceradenovec) – treatment for brain cancer (malignant glioma). *Ark Therapeutics: A specialist healthcare group focused on vascular disease and cancer, two of the largest therapeutic markets in the world.* www.arktherapeutics.com/main/research_development.php?content=products_cerepro.

Chapter 6

Looking backward: a model of value for translational trials

Introduction

In 2000, *Lancet* published results of a phase 1 study testing augmerosen in volunteers receiving the standard chemotherapy dacarbazine for melanoma. Augmerosen belongs to a class of genetic therapies – called antisense agents – that work by blocking expression of specific genes. In this case, the target was BCL-2, which prevents cells that have sustained DNA damage from undergoing "cell suicide." The idea of the study, then, was to use augmerosen to release the genetic "brake" on cell suicide so that tumor cells would die after chemotherapy.

The study results were favorable: augmerosen combined with the chemotherapy drug dacarbazine proved safe even at the highest dose tested, with no volunteers developing life-threatening and/or unanticipated toxicities. In addition, six of the fourteen volunteers showed tumor shrinkage, and for two others, tumors stopped growing. In one figure, numerous thick, pigmented growths crowd a volunteer's forearm before the start of treatment; after, the tumors are reduced to a series of small moles. Whereas patients treated for advanced melanoma might typically live another five months, these study volunteers survived a median of twelve.[1]

The study was valuable in the way phase 1 studies are typically valuable. According to one authoritative source, "the primary purposes of classic phase I studies are to investigate toxicity of organ systems involved, establish an optimal biological dose, estimate pharmacokinetics, and assess tolerability and feasibility of the treatment. Secondary purposes are to assess evidence for efficacy, investigate the relation between pharmacokinetics and pharmacodynamics of the drug, and targeting. Not all these can be met completely in any phase I trial."[2] Thus, essential safety information was collected over a range of doses, enabling the design of subsequent controlled studies of efficacy.[3,4] The authors showed evidence consistent with a decrease in BCL-2 expression and, when combined with a chemotherapy drug, enhanced cell suicide. A few years later, several follow-up phase 2 trials were launched on the basis of these results.

But was the study valuable enough? Many would answer yes: all that can be asked of phase 1 studies is that they provide enough information to plan a phase 2 study. But some might find the study wanting. The authors did not report pharmacokinetic data. They showed that expression of the target gene – BCL-2 – was appropriately down-regulated, triggering cell suicide in tumor tissue. But the data were from one patient only. How can we be assured that the investigators did not cherry pick their data? And even if these data were representative, might not the responses have been caused by an alternative mechanism? From this standpoint, the *Lancet* study exemplifies the low scientific standards that some believe prevail in early-phase clinical research.

In what follows, I examine what makes early phase 1 clinical trials socially valuable. I critique a mainstream framework for value, which casts phase 1 studies in a supporting relationship with phase 2 trials. Instead, I propose a model that views phase 1 translational clinical studies as Janus-like endeavors poised to stimulate preclinical research as well as further clinical development. I also examine whether factors more distant to the research process – an agent's likely cost, for example – should alter the appraisal of phase 1 study value.

The dark matter of research ethics

All influential codes of human clinical research ethics stress that human experiments are permissible only insofar as they produce socially valuable information. For example, the Nuremberg code states that experiments "… should be such as to yield fruitful results for the good of society;" elsewhere, the code states that "the degree of risk to be taken should never exceed that determined by the humanitarian importance of the problem to be solved by the experiment."[5] *

But what do these policies intend by "good of society" and "humanitarian importance," and how might they be assessed? Medicine has evolved sophisticated metrics to determine the health and economic value of existing therapeutic and diagnostic interventions. Through evidence-based-medicine and pharmacoeconomics, it has also produced somewhat cruder ways of ranking the utility of completed clinical studies. Nevertheless, how investigators, policy-makers, and ethics committees should *prospectively* assess a study's value remains virtually unaddressed.

Astrophysicists tell us that 96 percent of the mass and energy in the universe is "dark." Yet in spite of their abundance, neither dark matter nor dark energy is directly observable. In many respects, value represents a kind of "dark matter" for biomedical research ethics: it constitutes an essential and abundant "mass" that counterbalances risk for any human experiment. Ethical decisions around human clinical studies always hinge in part on a projection of a study's value. But research ethics has yet to firmly grasp this elusive entity. Despite its ubiquity as a justification for clinical research, a search of the scholarly literature on value turns up only a handful of conceptual papers amid thousands on consent, risk, inducement, and privacy.†

The question of how value should be defined and assessed is complicated enough in the context of controlled clinical trials. Phase 3 studies are frequently criticized for their choice of comparator, subject selection,[7] planning for subgroup analysis, adverse event reporting,[8,9] and choice of endpoints.[10] One major axis of dispute concerning value in randomized controlled trials concerns their epistemic orientation. Should trials aim at determining how and why a drug works? Such "explanatory" studies try to maximize the likelihood that therapeutic

* Similar sentiments are expressed in other codes as well. Paragraph 18 of the Declaration of Helsinki, for example, states: "Medical research involving human subjects should only be conducted if the importance of the objective outweighs the inherent risks and burdens to the subject."

† Several provisos are in order for this broad claim. First, there is a kind of "shadow" literature on ethics value in the biostatistics and epidemiology literature. That is, to a large degree, research ethics has ceded a philosophy of value to biostatisticians. Second, part of the reason why value has been neglected as a subject for sustained ethical analysis must surely be the intractability of the problem. Whereas most research ethics centers around questions of means, value forces a confrontation with ends. In his book on the sociology of debates over gene transfer, John Evans argued that, through the 1970s and 1980s, ends-oriented ethical analysis was eclipsed by means-oriented ethical analysis. Without pronouncing here on Evans's account, the reticence of research ethics on questions of value is consistent with Evans's thesis.[6]

effects will be observed by controlling the setting and circumstances in which volunteers receive experimental interventions. For example, explanatory studies tend to avoid enrolling persons with co-morbidities, and exclude from analysis data on volunteers who drop out during a study. Alternatively, trials might be designed to determine whether a drug will be useful in the types of clinical settings in which the drug will be used. Such "pragmatic" designs tend to use broader eligibility criteria for volunteers, and allow the drug to be given in different types of clinic. Pragmatic trial design will generally perform hypothesis testing on the basis of "intention to treat" (that is, data on volunteers who drop out during the study are nevertheless included as if they had continued enrollment).[11,12]

How such debates are resolved has important implications for future patients, because controlled clinical trials shape the knowledge environment for medical practice. But the consequences for trial volunteers are more circumscribed, because clinical equipoise establishes a limit on the risks they are asked to endure. For example, a 2001 review that found that 40 percent of randomized controlled trials involving acute coronary syndromes exclude persons older than age 75, even though such persons account for 37 percent of those with myocardial infarction. One of the reasons elderly volunteers are excluded is that co-morbidities in aged patients greatly complicate the analysis of safety data.[9] Such "explanatory" design has major consequences for the care of elderly patients, because it effectively deprives them of "evidence-based care."[13] However, if these trials fulfill the care-based conditions of clinical equipoise at their inception, no volunteers enrolled in any of these biased studies knowingly received therapy that was demonstrably inferior than standard of care.

How value is defined for phase 1 trials of novel interventions has limited implications for current patients, because studies are not directly aimed at informing current clinical practice. But how policy-makers and investigators conceive the value of phase 1 studies has important consequences for research subjects. Recall that Chapter 5 argued that translational phase 1 trials involving novel interventions (by contrast with phase 1 trials involving combinations of validated drugs, or using reasonably well-characterized drugs for new indications) are generally not justifiable in therapeutic terms. If this claim is valid, then ethical acceptability depends entirely on an appraisal of a study's value.[9]

The regulatory model of value

The first thing to point out about phase 1 studies is that they serve different ends than phase 3 studies. Controlled clinical trials are generally considered valuable if, upon completion, there is a reasonable probability that findings will alter clinical practice. In order to achieve this objective, a study must be adequately powered to test a specific, explicitly formulated hypothesis. However, phase 1 studies are generally inadequately powered to influence clinical practice directly, do not typically involve control groups, and rarely specify clear hypotheses towards which the study is directed.

Instead, the value of a phase 1 trial is typically judged by whether it can "reliably determin[e] whether a new therapy shows at least some promise of benefit, and ... explicitly aim ... at guiding a definitive phase 3 trial that will be adequately powered to make a reliable treatment comparison."[14] The augmerosen trial with which this chapter opened provides a clear example of how phase 1 studies can attain this type of value. The study reduced uncertainty about the clinical properties of augmerosen combined with chemotherapy by producing information about safety. It provided evidence for designing more statistically meaningful tests of efficacy by identifying appropriate doses for future studies. And it informed decisions about whether to proceed into phase 2 testing by collecting evidence of biological effect.

Such a formulation of value for phase 1 studies is what I will call the "regulatory model" of phase 1 studies, since it corresponds with the FDA definition of phase 1.[‡] Because this formulation views phase 1 studies in terms of their ability to promote progression of interventions through to various clinical testing phases to licensure, I call this "progressive value."

The translational model of value

A series of hemophilia clinical studies performed at the Children's Hospital of Philadelphia illustrates a different way that phase 1 trials sometimes contribute knowledge. An initial phase 1 study had tested the effects of administering AAV2 vectors carrying a factor IX transgene to muscle.[16] Though serum tests indicated expression of coagulant factor IX, the researchers projected that at least 500 muscular injections would be required to achieve therapeutic levels. The researchers also noted that expression of the therapeutic transgene was considerably lower than anticipated on the basis of experiments in dogs. Seeking a more practical approach, the researchers designed a protocol that administered identical vector to the livers of seven volunteers. This time, serum coagulant factor in two volunteers receiving the largest doses edged to therapeutic quantities, but then diminished to undetectable levels.[17] Follow-up testing indicated that factor levels declined in lockstep with an immune reaction against vector. At least two hypotheses might explain this turn of events.[18] First, vector might be adhering to the outside of liver cells, tempting the immune system to mount a reaction against the transformed cells. If so, a transient regimen of immune-suppressing drugs should be enough to enable liver cells to absorb the capsid proteins and enable long-term survival of coagulant factor-expressing liver cells. Second, experiments in a different laboratory suggested that AAV2 might have a propensity for being taken up by dendritic cells, a type of blood cell that is responsible for amplifying immune reactions. If so, different AAV vector serotypes – say, serotype 8 – might overcome this problem. Researchers developed protocols structured around testing each hypothesis.

Thus far, clinical achievements in these studies are modest and have not led to phase 2 testing. But they have produced unexpected findings about human immune responses to AAV vectors, and the differences between dog and human models. They also improved knowledge about the biodistribution of AAV vectors. They identified immunological barriers for AAV-based gene transfer, and led to the formulation of two discrete, testable hypotheses. Both of these are currently being followed up in preclinical and phase 1 studies. A raft of studies involving AAV vectors for other diseases is using a short course of immune suppression. In short, though the outcome was null if judged on the basis of progressive value, the study was valuable for a number of other medical and scientific reasons.

In 1995, NIH director Harold Varmus assigned hematologist Stuart Orkin and medical geneticist Arno Motulsky to chair a panel assessing the status of human gene-transfer research. The committee's report faulted human gene-transfer research for "insufficient

---

[‡] 21 CFR 312.21 defines phase 1 studies as "designed to determine the metabolism and pharmacologic actions of the drug in humans, the side effects associated with increasing doses, and, if possible, to gain early evidence on effectiveness. During Phase 1, sufficient information about the drug's pharmacokinetics and pharmacological effects should be obtained to permit the design of well-controlled, scientifically valid, Phase 2 studies." The FDA definition also includes another paragraph stating "Phase 1 studies also include studies of drug metabolism, structure–activity relationships, and mechanism of action in humans, as well as studies in which investigational drugs are used as research tools to explore biological phenomena or disease processes."[15]

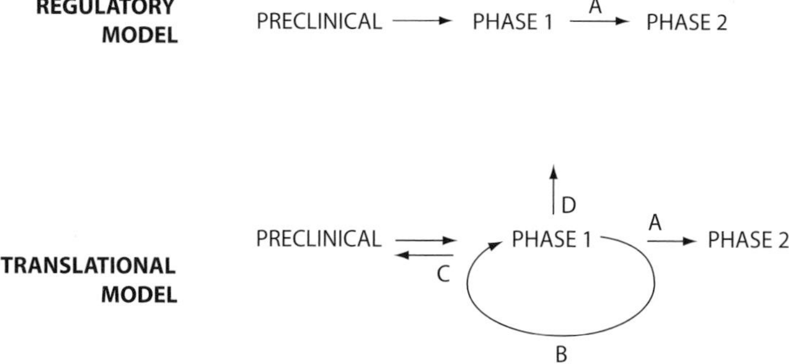

Figure 6.1 Models of value for phase 1 studies. "A" refers to progressive value; "B" to iterative; "C" to reciprocal; and "D" to collateral.

attention to research design, poorly defined molecular and clinical endpoints, and lack of rigor." The report's passages on study design bear quoting:

> Clinical studies represent not only practical implementation of basic discoveries, but also critical experiments which refine and define new questions to be addressed by non-clinical investigation .... well-designed studies greatly increase the information that may be extracted from careful clinical experiments involving only a few patients.... [C]linical gene therapy studies reveal problems and raise questions that cannot be otherwise anticipated .... Gene therapy clinical research may provide insights into fundamental disease pathology that may direct subsequent treatment approaches ... Reciprocal and synergistic relationships between clinical studies and basic research may emerge from initial clinical gene transfer studies ... For this and other fields of clinical investigation to succeed, high standards of experimental design and robust methods for evaluating clinical outcomes are needed. All studies should define molecular, biochemical, and quantitative clinical endpoints. *They also need to address specific hypotheses, enabling investigators to interpret negative as well as positive findings* [emphasis added].[19§]

This passage posits an alternative to the regulatory model of value for phase 1 studies of novel interventions, what I will call the "translational model" (similar sentiments have been expressed by a wide variety of translational researchers).[21,22] Specifically, trials can produce useful knowledge by motivating further preclinical studies of an intervention ("reciprocal value"), by prompting modification of human translational trials of a particular agent ("iterative value"), or by informing other areas of loosely related research practice ("collateral value") (see Figure 6.1). Thus, the liver–AAV–hemophilia study produced reciprocal value by stimulating further refinements in AAV vectors (for example, see Wu *et al.*[23]), studies of how AAV interacts with the immune system,[24] and preclinical studies leading to a new human trial.[25] The facts that the intramuscular trial led to the hepatic study, and that the hepatic study has itself led to two different hepatic trials in hemophilia, show how the study achieved iterative value. The study achieved collateral value through a better characterization of AAV safety (for example, we now better understand the predictive value of rabbits for AAV biodistribution), and more fundamental work on mechanisms of the human immune response to AAV (see, for example Zaiss *et al.*[26]).

[§] This is not the first time such a view has been articulated with respect to gene transfer. Indeed, Clifford Grobstein and Michael Flower said as much as early as 1984![20]

Biostatistician Steven Piantadosi formalized the aims and method of translational trials as "guid[ing] further experiments in the laboratory or clinic, inform[ing] treatment modifications, and validat[ing] the target... translational trials imply circularity between the clinic and the laboratory, with continued experimentation as the primary immediate objective. Many therapeutic ideas will prove useless or not feasible during this cycle. The laboratory-clinic iteration may eventually beget the familiar linear development of a new therapy, perhaps after numerous false starts."[27]

Favoring the translational model

That the classic regulatory model of value prevails in popular, commercial, and medical discourses is clear in the language of drug development and clinical research. In 2004, the FDA launched its "Critical Paths" initiative aimed at improving the translation of medical innovations into clinical practice. As a first step, the FDA issued a white paper providing at an "analysis of the *pipeline problem* – the recent slowdown, instead of the expected acceleration, in innovative medical therapies reaching patients."[28] The pipeline metaphor – like the "bench-to-bedside" expression that is ubiquitous in drug development – suggests a unidirectional, linear flow of drugs and information. Another term in drug development is "attrition:" the abandonment of new chemical entities owing to concerns about safety and efficacy. From a commercial standpoint, any failure to advance a product to licensure is an unmitigated waste of resources. But unlike attrition in war and college admissions, failure to progress to later-stage clinical testing can lead to important physiological and pharmacological insights.[29]

Descriptions of phase 1 studies in the patient literature also suggest a progressive and sequential orientation. The US National Cancer Institute (NCI) Cancer Therapy Evaluation Program (CTEP) defines the objective of Phase 1 trials as "determin[ing] a safe dose for Phase 2 trials and defin[ing] acute effects on normal tissues."[30] Elsewhere, the NCI describes phase 1 studies as "the first step in testing a new approach in people ... the highest dose with an acceptable level of side effects is determined to be appropriate for further testing."[31] Similarly, the American Cancer Society's leaflet on clinical trials states "The main reasons for doing phase I studies are to determine the highest dose of the new treatment that can be given safely (without serious side effects) and to decide on the best way to give the new treatment... If a new treatment is found to be reasonably safe in phase I clinical trials, the treatment can then be tested for effectiveness in a phase II clinical trial."[32] A UK leaflet on cancer clinical trials states: "the aim of a phase 1 trial is to find out if a new drug or treatment is safe in people and how much of it should be given."[33] Similar definitions are recapitulated in other influential descriptions of phase 1 studies.[34]

Finally, the regulatory model of phase 1 value is visible in the way studies are branded as failures or successes. A *Nature Medicine* press release states "[f]ollowing successful phase I trials of a form of gene therapy... researchers... now report promising Phase II results...",[35] and researchers refer to "successful phase I clinical trials" of gene-transfer agents that "attest to their safety in humans."[36] A Huntington's Disease advocacy website eagerly reports on "[s]uccessful Phase 1 Gene Therapy Trial" involving Parkinson's disease.[37]

The classic regulatory model has generally worked well for cytotoxic cancer drugs, where the basic strategy – preferential toxicity for dividing cells – is fairly clear, and the therapeutic approach is to deliver a maximum tolerated dose. But is it an appropriate way to think about translational clinical studies? On what ethical basis might Orkin and Motulsky ground their harsh criticisms of the first 100 gene-transfer clinical studies? What is the ethical case for favoring the translational model when novel agents are first tested in human beings?

To begin with, the two models are not mutually exclusive: a trial can pursue the primary objective of establishing dosage for phase 2 testing while testing molecular and physiological response. Nevertheless, the models have different implications for design, risk assessment, and informed consent. The translational model, for example, would encourage more translational studies like exploratory trials where molecular response – rather than safety and dosage – is the primary endpoint. It would also tend to favor diseases where translational studies are feasible. For example, cancers involving the skin, head, or neck provide relatively accessible tissue for biopsies, and hence more amenable to correlative studies of biomarkers than are solid-organ tumors. Similarly, it is probably no coincidence that many of the most informative and clinically successful gene-transfer studies have involved blood disorders, for which tissues are relatively easy to procure for characterization (either that, or hematologists are a little smarter than the rest of us).

The translational model will also tend to have a more stringent and elaborate assessment of risk and benefit in phase 1 studies, viewing the correlative studies as the "meat" of the protocol rather than as the "gravy." For example, reviewing phase 1 studies with an eye towards the translational model would require that the committee appraise the predictive value of biomarkers, the sample size and powering of correlative studies, and the possible deposition of tissue specimens in biobanks. In contrast, all that the classic regulatory model requires for assessing risk is application of well-established cohort sizes and dose-escalation regimes.

As for consent, it is probably a safe bet that trial participants view phase 1 value largely in progressive terms (though, to my knowledge, how patient volunteers conceptualize the goals, value, and proximate application of clinical research findings has yet to be adequately explored empirically). If the translational model is more appropriate, phase 1 trials need to be presented to volunteers as quasi-scientific explorations rather than as truck stops on the highway toward clinical application.

Three related arguments favor the translational model when testing novel interventions in human beings. First, it is increasingly apparent that the model of scientific and technological advance articulated by Vannevar Bush (that is, the notion that innovation proceeds unidirectionally from basic science to application; see Chapter 2) is less and less tenable. In his book *Pasteur's Quadrant* former University of Michigan dean Donald Stokes describes numerous instances where major scientific discoveries flowed from applied research (examples include Langmuir's discoveries in surface chemistry, which originated in work in electronics; within medicine, examples include the discovery of triplet expansion in genetic disease followed from clinical observation.[38])[39, 40] Within first-in-human trials, recent autopsy data on volunteers participating in fetal tissue transplant trials for Parkinson's disease have led to major new pathophysiological insights.[41–44] The translational model is more consistent with this iterative view of discovery and innovation.

Second, the translational model seems to more accurately reflect the way that translational trials accrue value. With David McLaughlin, I examined patterns of citation of phase 1 studies under the assumption that citation numbers and characteristics reflect some aspect of how peer researchers value a study (for example, highly valued studies will tend to be cited more frequently than those that are less valued).[¶] We then collected a large sample of phase 1 studies

[¶] We restricted our search to phase 1 studies involving malignant glioma to reduce variation in citation patterns due to differences in how much basic and applied research is performed for different diseases. We specifically selected malignant glioma because this disease category affords one of the largest samples of gene-transfer phase 1 studies.

and divided them into three categories: non-novel agents (i.e. those involving agents that were already licensed for some indication at the time the trial was published), novel agents (i.e. those involving at least one agent where the product was not yet licensed for any indication), and very novel agents (i.e. those that involved gene-transfer strategies). Using ISI Web of Science, we collected subsequent articles that cited these phase 1 trials. Citing articles were then divided into clinical trials, animal studies or in vitro experiments, and physico-chemical studies. Clinical trials were further classified according to their phase. Our working thesis was that clinical trials that cite phase 1 studies reflect instances where citing authors value the iterative or progressive information of the trial. Animal studies citing human trials reflect instances of reciprocal value. In vitro and physico-chemical studies reflected still more basic forms of reciprocal value.

Not surprisingly, we found a consistent relationship between the novelty of the intervention and the number of citations to the phase 1 study (e.g. licensed-agent trials received on average 2.5 citations per year, as compared with gene-transfer agent trials, which were cited 16 times per year). Within gene transfer, studies using new vectors received far more citations than those using more familiar ones. This simply reflects the fact that biomedical researchers tend to value the information contained in studies of novel agents more than they do the information contained in studies of well-characterized drugs. More pertinent for our discussion of what kinds of information researchers draw from phase 1 studies, we found a relationship between novelty of interventions and the proportion of citing articles that were animal or in vitro studies. In contrast, less novel intervention trials were proportionately more likely to be cited by phase 2 clinical trials. This relationship held even when we divided citing articles according to the reasons they cited the phase 1 study. We inferred from these results that the information gathered from novel-agent studies is much more likely to be incorporated into preclinical studies than new phase 1 or phase 2 studies. The reverse is true for studies of already licensed agents.**

This study of citation patterns provides some empirical support for my third defense of the translational model: studies designed for iterative, reciprocal, or lateral value are more likely to meaningfully inform clinical and scientific practice than those that are not. As noted in previous chapters, gene-transfer interventions are complex and composite, and the human body has evolved poorly understood mechanisms of resisting them. Uncertainty in such studies is virtually unbounded, and the reasons an intervention might unsuccessfully translate are myriad. Under such conditions, a large majority of phase 1 studies will either fail to justify phase 2 testing, or at least not lead directly to licensed interventions (see Box 6.1).

With the odds stacked so heavily against advancement to later-stage clinical testing and/or eventual licensure, phase 1 trials structured entirely around a progressive notion of value

** Our studies of citation patterns have many limitations. Citations, for example, are imperfect indicators of how peer communities value published articles; they are also prone to a number of biases (for example, European studies tend to be cited at a lower frequency than US studies[45]). Second, our classification of novelty is admittedly crude: a range of novelty is likely to exist within each category. Third, there is no assurance that these citation patterns would be the same for other diseases. Nothwithstanding these concerns, these data are consistent with the notion that novel intervention studies tend to inform preclinical and basic research more than they do further clinical development, and that the reverse is true for interventions that are already licensed.

are unlikely to deliver on their scientific and social benefits, and hence redeem the (therapeutically unjustified) risks and burdens presented to volunteers. Though they might have a modicum of social utility in that they help to rule out a particular intervention applied in a particular way, unsystematic shots in the dark do little to meaningfully inform biomedical practice. Consider how this contrasts with randomized controlled trials. Several analyses of publicly funded studies (where publication bias is improbable) found that, on balance, new interventions out perform standard of care in only 50 percent of trials.[46–48] This probability of yielding information that could modify medical practice is the highest that could be expected – or desired (any higher probability would raise ethical concerns that persons on the "standard treatment" arm systematically receive inferior care).

In summary, few novel intervention trials lead seamlessly to phase 2 studies and eventual licensure. For early-phase studies to significantly inform biomedical practice and hence redeem the risks they present to volunteers, they need to be structured such that negative findings can be interpreted meaningfully and lead to new lines of experimentation. This is best accomplished by designing human translational studies that can be anticipated to capture reciprocal, iterative, and collateral value should progressive value not materialize.

Box 6.1 Success rates for advancing new agents to phase 2 testing

What is the frequency with which novel compounds and/or biologics in first-in-human trials advance into phase 2 testing? What is the frequency with which such compounds ultimately advance into phase 3 studies and/or licensure? To this author's knowledge, there are no good studies addressing these questions.

One study of new chemical entities that enter clinical development indicated that only 11 percent will translate to licensure successfully (the figure is as high as 20 percent for cardiovascular interventions, and as low as 5 and 8 percent for cancer and central nervous system drugs, respectively.)[49] As many as 60 percent of phase 1 studies lead to phase 2 studies. However, these figures are derived from the ten largest pharmaceutical companies, and thus might not represent scientifically intensive phase 1 trials of the sort pursued in academia and the biotechnology industry. In addition, these figures exclude biologics. They thus probably provide "best-case" success rates for novel compounds.

The NIH maintains a database of all gene-transfer trials pursued at institutions receiving funding for recombinant DNA research. A crude search of this database shows that 148 phase 2 protocols have been submitted; this corresponds to 22 percent of all phase 1 protocols. The Wiley database of gene-transfer trials worldwide lists 216 phase 2 studies, which corresponds with 26 percent of phase 1 studies (phase 1/2 studies were excluded from this calculation). One might infer, then, that approximately one in five agents entering phase 1 study reach phase 2, and (as of yet), none reach licensure. Again, what is confusing about these figures is that, even for gene transfer, only a minority of phase 1 studies are first in human, many approaches are not altogether novel, and multiple phase 2 studies are sometimes run on a single agent. Indeed, there are some examples within the NIH database where phase 2 studies are listed without corresponding phase 1 studies. Once again, this likely results in an overestimate of the frequency with which novel agents reach phase 2.

Because of fairly high rates of nonpublication for phase 1 studies (in the field of cancer approximately 33 percent are not published after 7.5 years[50]), literature searches might not provide a reliable measure. An alternative approach would be to examine how many novel compounds entering clinical testing in the NCI's CTEP program reach later phases of testing. Until studies like this are performed, we are stuck with very crude (and likely, optimistic) forecasts of clinical advancement.

Applying the translational model

Epidemiologist Steven Piantadosi offers one of the few formal justifications and structures for translational clinical studies. His account is descriptive and not intended as a critique of current phase 1 study designs. Nevertheless, he lays out ten defining characteristics for translational clinical trials that might serve as guideposts for designing first-in-human phase 1 protocols of novel interventions that have a reasonable prospect of achieving iterative or reciprocal value.[‡‡] I take Piantadosi's ten characteristics as reducing to five elements:

(1) *Solid preclinical evidence base*: Translational trials should be "predicated on promising preclinical developments, which create a need to evaluate the new treatment in human subjects." The basis for this requirement is that negative studies are uninformative where preclinical evidence does not strongly support success in human beings. A positive result in animals, followed by a negative result in human beings, can lead to further exploration of the basis of this difference, as well as to the development of strategies to circumvent human barriers to success. This theme will be explored at greater length in Chapter 7.

(2) *Flexible intervention*: "The treatment and/or its algorithm is changeable, perhaps following additional laboratory experiments." This criterion preserves a potential for subsequent phase 1 studies to be designed using findings in the translational study. In practice, few gene-transfer protocols are likely to present challenges for fulfilling this requirement. If expression of a transgene is insufficient, different promoters or gene constructs can be substituted; if vectors target inappropriate tissues, different delivery systems or envelope proteins can be tried.

(3) *Explicit description*: "The treatment and its method of evaluation are formally and fully specified in a written protocol." Though this is necessary for any clinical trial, it will often present particular challenges for novel and complex interventions like gene- and cell-based technologies, where vector or cell compositions often vary along several dimensions, methods and standards for characterizing interventions are not well developed, and surgical techniques for delivery limit specification. The challenges of specifying agent have been discussed in Chapter 3.

(4) *Measurement of target effects*: "The evaluation relies on one or more biological targets or outcomes that provide definitive evidence of mechanistic effect within the working paradigm of disease and treatment." This provides an opportunity to confirm that the agent has the desired mechanistic effects, and if not, provides a starting point for identifying reasons why an agent behaved differently in preclinical models. To ensure validity of target–effect measurements, two conditions should be fulfilled. First, "large effects on the target are sought." This improves the strength of causal inference when effects are observed. Second, protocols should have an "explicit, unambiguous definition of 'lack of effect.'" This helps reduce opportunities for bias in measurement and interpretation.

(5) *Contingency plan*: "The study protocol specifies the next experimental step(s) to be taken for any possible outcome of the trial. Consequently, regardless of results, the trial will be informative with respect to a future experiment." In other words, a translational trial

[‡‡] From here on, this chapter will not discuss collateral value, which is often difficult to plan for. Nevertheless, Chapter 3 and what follows in this chapter outline a number of practices that enhance the likelihood of capturing collateral value with respect to safety. These include detailed characterization of experimental intervention (using standardized assays), full reporting of subclinical observations, and biobanking of tissues for later correlative studies.

should have an explicit and prospective plan for capturing and applying iterative and/or reciprocal value.

These five elements might be applied to novel intervention trials as follows. For any protocol to return iterative or reciprocal value, each of these elements must be present. Thus, if we accept the claim above that all first-in-human trials involving novel interventions should be valued along the translational model, all such protocols should be expected to satisfy a threshold condition for each of the five elements, or at least explain why an element is inapplicable or expendable.

Furthermore, note that each of the elements can vary with stringency. That is, reviewers and policy-makers might in some circumstances demand stronger preclinical data, greater flexibility for an intervention, more detailed description, more numerous and better-quality target effect measurements, and a more developed contingency plan. In many circumstances, the informative value of a translational protocol will correlate with the degree to which each element is addressed. I would suggest that two factors should determine the stringency of standards against which each of the five elements should be measured. The first is risk: trials presenting greater risk should be expected to return greater value. Therefore, riskier protocols should have a more developed plan for addressing each element. The second is complexity: where interventions are compositionally or functionally complex, the need for stringent fulfillment of these elements will be greater. This is because "negative results" will tend to be less informative where there are multiple points in the causal chain at which the intervention can fail. For example, a novel cancer vector might work by selectively entering tumor cells, expressing a "cell suicide" (apoptosis)-inducing transgene, triggering apoptosis, and causing destruction of neighboring tumor tissue through what gene-transfer researchers are calling the "bystander effect."[§§] As described here, the biological response will depend on at least four broad causal steps.[¶¶] Imagine a trial where this vector fails to cause tumor shrinkage in a group of cancer patients. Fortunately, the investigator anticipated this outcome and performed an analysis to determine whether she could detect apoptotic cells. Her results are negative. Though this finding is helpful, it does not tell us whether this is a consequence of poor transfection, low transgene expression, or failure of the transgene to cause the desired biological effect. To nail down the cause of product failure, reliable measurements of transfection, transgene expression, and arguably, transgene product are needed.[***] Thus, such trials

§§ The causes of the bystander effect are not well understood. One prominent theory is that death of transfected tumor tissue provokes an immune recognition of abnormal tissues nearby.

¶¶ Of course, this is a pragmatic statement. The "four causal steps" have no ontological status, as each depends on a number of substeps which, should they come to be seen as significant, could easily add yet another step. For example, the bystander effect might consist of several distinct steps. Nevertheless, scientists often break causal processes into distinct steps, and I am merely pointing out here that, for the purposes of trial design, forethought about these various steps is desirable.

*** Consider also an alternative scenario, in which a strong tumor response is observed, and the investigator observed signs of apoptosis. It remains to be determined whether the transgene caused the apoptosis, or whether programmed cell death is simply the consequence of viral transduction.Oncology gene-transfer researchers Steven Albelda and Daniel Sterman suggest transfection and transgene expression as standard assays that – unless they impose unacceptable risk for volunteers – should be expected for gene transfer trials: "Given our very limited knowledge about the efficiency of gene transfer in humans, gene transfer trials should try to incorporate methods to assess transgene expression. Without knowing if a vector has effectively transduced the intended cells and produced the transgene in clinically useful amounts, it is very difficult to predict if the approach chosen will work as hypothesized and how the next phase I or phase II trial can be designed. This information is especially important given that vectors such

should collect information at multiple points in the causal chain from agent administration to clinical effect.[†††]

Last, for translational studies to return value without imposing greater burdens for volunteers, researchers should generally favor initial testing of novel interventions in diseases for which tissues are relatively accessible and pathology well understood. This represents somewhat of a reversal of the clinically directed approach to research, in which interventions are chosen because of the properties of a particular disease. Instead, the translational approach selects diseases for initial tests of an intervention at least in part on the basis of amenability to correlative and molecular studies. This logic might explain why a disproportionate volume of early gene-transfer trials targeted head and neck cancers. It also relates to the reason so many gene-transfer studies have involved very rare immunological diseases and hemophilia: these disorders continue to be seen as the most likely systems for attaining progressive value. The last twenty years reveal just how elusive progressive value can be for novel interventions. Nevertheless, accessibility of blood cells, marrow, and blood sera make these diseases attractive for informative molecular studies that can produce reciprocal, iterative, and collateral value.

Objections to the translational model

Practical limitations

The foregoing invites several criticisms. The types of correlative study necessary to fulfill the translational model are costly. Precise amounts might vary between 10 and 15 percent of the total budget for a study if tissue access is difficult (say, it involves general anesthesia); the cost will be significantly lower if tissue is relatively accessible. In one study, the per-subject costs increased from $7500 to $15 000 when biomarkers were included.[51] To a university or biotechnology firm, correlative studies might not be financially viable. Moreover, small biotechnology firms are sometimes said to avoid correlative studies because of concerns that a drug might not work by the mechanism originally postulated. The Genasense® studies provide one of many instances where the mechanism of a novel drug turned out to be quite different than predicted in preclinical studies. According to one cancer researcher I spoke with, discordant mechanistic studies are a liability for biotechnology companies, because they make it difficult to "keep their stories straight" for investors. As such, some firms would "rather not know."[‡‡‡]

However, my framework might be construed as ethically optimal rather than as obligatory, in which case ethics committees and others might implore rather than require researchers to pursue correlative studies. This might represent a pragmatic compromise in today's political environment. If it is, the onus shifts to ethics committees to nudge investigators where they can, and to right the consent process by making sure that volunteers are informed that the study they are entering meets minimally adequate standards of scientific value, but no more.

---

as adenoviruses have transgene-independent biologic effects that may not be readily discernible in the absence of "empty-vector" control groups."[21]

[†††] This chapter does not explore how frequently first-in-human studies perform the types of correlative study advocated here. One well-known gene-transfer researcher, when asked, informed me that about 50 percent of first-in-human cancer gene-transfer studies test transfection and gene expression – and that most of the studies that don't are industry funded. It would make a helpful contribution to the literature to explore this question more systematically.

[‡‡‡] It is very difficult to verify this claim. Some translational researchers I spoke with find this statement "overly cynical."

I nevertheless worry that this pragmatic response concedes too much. It would be highly unusual for study sponsors to refuse a request from an IRB or drug regulatory agency for further toxicology studies on the grounds of expense: adequate toxicology studies are considered a normal cost of doing business in clinical research. I cannot think of any reason why adequate value should not similarly be considered a standard cost for performing translational clinical research. In addition, the vast majority of translational clinical research on novel agents takes place at university medical centers, which provide the equipment, personnel, and patient pool to enable such studies. Notwithstanding the ascension of the entrepreneurial academy, the foremost commitments of research universities remain education and the production of knowledge. Universities cannot hold firm to these twin commitments while providing a setting and services for mediocre experiments.[§§§]

Limiting opportunities for unexpected discoveries

Another objection to the translational model – and my tracing of its implications – is that it is excessively conservative: it tends to favor methodical accretion of findings rather than stabbing into the unknown. This criticism might take several forms. Perhaps the least persuasive is the suggestion that scientific progress is serendipitous, and that my proposal does not provide sanctuary for unexpected discovery. This reasoning seems muddled. By definition, serendipity refers to accidental discovery in the course of looking for something unrelated. As such, serendipity is unplanned; it also does not follow that serendipitous discovery cannot emerge with similar frequency from conservative research design, or that serendipitous discovery is more common when research is less structured.[¶¶¶]

A more compelling criticism is that my analysis would appear to exclude high-pay-off research initiatives that break from convention. Outside of medical research, there are many examples of outlandish ideas that were nurtured into transformational technologies. One is the Internet. Another is the stealth bomber. In 1958, the US military established an agency, the Defense Advanced Research Project Agency (DARPA), to incubate "outside-the-box" concepts into applications. The DARPA model was emulated by the NCI with the creation of the Unconventional Innovations Program, which sought to develop diagnostic technologies.[52] A 2007 National Academy of Sciences report on US competitiveness in research and development urged federal research agencies to "allocate at least 8% of the budgets to ... be focused on catalyzing high-risk, high-payoff research of the type that often suffers in today's increasingly risk-averse environment."[53] And the NIH Roadmap initiative established its "Pioneer Award" program to allow research that is unusually innovative and reward "profound leaps of intuition" over "timid, incremental steps."[54] Clearly, contemporary research policy places value on eccentric research. Why not allow some space for wacky studies involving human volunteers?

There are several concerns about extending the DARPA model to human experimentation. Agencies like the DARPA are driven by an actuarial logic in which resources are invested

[§§§] I can concede that there might be some instances where correlative studies are prohibitory. It might be appropriate to preserve a space within ethics review and funding for investigators to rebut the presumption that their protocols contain correlative components.

[¶¶¶] I suspect that, if anything, the translational model proposed here would tend to favor serendipitous discovery to a greater extent than the regulatory model: because the former favors more collection of tissues and data types, it samples a wider variety of phenomena than a regulatory-directed study, which would simply collect information on pharmacokinetics and toxicity. This may provide more opportunity for unexpected discoveries.

in numerous dicey projects on the premise that one large pay-off will underwrite those that fail. In order to ensure that this research funding is wisely apportioned among risky projects, DARPA must constantly evaluate different programs, pruning ones that appear unpromising, seeding new projects, and bolstering flourishing programs. As such, an agency like the DARPA assesses the merits of one risky research undertaking in part on the basis of its portfolio of other risky ventures that are in varying stages of development. In contrast, IRBs review trial protocols one at a time. They are thus not in a position to determine whether a highly innovative, "high pay-off" trial has a favorable risk–benefit balance when viewed in the aggregate across a collection of other highly innovative studies.

Perhaps a deeper concern about extending the DARPA model to clinical research is that human beings – unlike research budgets – are not fungible. The logic behind DARPA-type programs is to bundle together a group of research proposals that, if viewed as individual projects, would represent poor investments. This approach creates problems when extended to human studies. Specifically, it requires that ethics committees allow studies that, when viewed alone, offer an unfavorable risk–societal benefit balance. But it is not consistent with research ethics that a favorable risk–benefit balance for volunteers entering a study be "purchased" by a collection of other protocols' highly favorable risk–benefit balances.

Physical limitations

Another important objection to the translational model of value presented in this chapter is that certain diseases might not be amenable to correlative studies. For example, gene-expression studies are difficult to justify ethically if they involve gene transfer to the brain. A strict requirement for translational studies would hamper human studies against neurodegenerative diseases. There are several clear responses to this argument. First, correlative studies might use imaging. Second, they might take place during post-mortem examinations, in which case what is needed is a well-developed plan to secure consent from volunteers and their families for post-mortem tissue collection. Third, circumstances where tissue cannot ethically be obtained because of burden and where there is no related disease affording more accessible tissues should count as exceptions with respect to the requirement for correlative studies.

Ethical limitations

A final concern about translational value is that correlative studies necessitate burdensome and occasionally risky procedures like serial biopsies or imaging studies. This asks more of the otherwise debilitated research volunteer. Perhaps even more troubling, most volunteers will be motivated to enter studies by the prospect of direct benefit. Expecting correlative studies from translational trials, some might argue, yokes scientific progress to the wishful thinking and desperation of study volunteers rather than their desire to promote research.

I acknowledge this tension exists. One way of reducing it is study design: as noted already, researchers should pick diseases where tissue is superficial. Alternatively, correlative studies might be piggy-backed on procedures. For example, scheduled operations can provide an opportunity to collect tissue that would otherwise be inaccessible. Another way of reducing the tension is a more thorough informed consent that emphasizes the role of study volunteers as partners in the research process rather than as a patients receiving care. Last, this objection should be viewed in the context of my argument that first-in-human trials are, for the purposes of ethical evaluation at least, non-therapeutic. The problem of yoking progress to wishful thinking is virtually constitutive of studies involving patients with few options, and correlative studies will usually account for a small portion of a study's total burden and risk.

The humanitarian model of value

Many of the above criticisms view my proposal for value assessment as restrictive. But there is another objection, one that might regard my position as too permissive because it takes a myopic view of value. According to this criticism, clinical research is valuable because it advances broader societal ends like improved healthcare; and healthcare is valuable only insofar as it leads to human flourishing. This claim finds its basis in various ethical codes and commentaries. For example, recall the Nuremberg Code's appeal to "the good of society" discussed earlier in this chapter.**** Other commentators argue that "research … in which the intervention could never be practically implemented even if effective is not valuable."[55–57] Any measure of a trial's value should therefore take into consideration the practical significance of the research.

The "humanitarian model" presents challenges for the translational model. A line of research that, if successful, would lead to an effective vaccine against HIV surely would have greater humanitarian significance than one aimed at developing wrinkle creams. As such, translational studies involving the former should, if necessary, tolerate greater risk than those involving the latter. But the translational model of value, focused as it is on the flow of knowledge rather than the ends towards which the knowledge is to be applied, does not provide any basis for this judgment. Because the humanitarian model raises serious questions for valuing early-phase clinical research of novel interventions like gene transfer, I will devote the remainder of this chapter towards exploring its implications.

A humanitarian framework for assessing a study's value might divide along three axes. One would be the severity of an unmet health need addressed by the research. However beneficial a society might consider treatments for erectile dysfunction, studies aimed at developing treatments for lethal and highly morbid conditions like primary immune deficiencies are, all else being equal, more valuable.

In one exploration of value in clinical research, David Casarett and colleagues offer two other axes for a humanitarian framework.[58] One is temporal: the results of some studies have immediate application, as when a phase 3 study shows a new drug to be more effective than standard treatment. The humanitarian value for other studies, like phase 1 trials, is postponed. All other things being equal, studies with immediate benefits are more valuable than those whose value is postponed, because the former have a greater likelihood of application.

The other axis proposed by Casarett *et al.* is population. Some studies will have narrow applicability to specific populations; an example would be a phase 3 trial involving interventions against schizophrenia. Other studies might have broader benefits: studies examining the properties of angiogenesis factors in human beings have implications for the treatment of such diverse conditions as cancer, age-related macular degeneration, limb ischemia, and wound healing. Still other studies might target populations deemed a social priority. For instance, on both humanitarian as well as economic grounds, one might argue that treatments aimed at pediatric disabilities are more valuable than those targeting the disabled elderly.

Given the diversity of gene-transfer interventions being pursued, only broad brushstrokes can be offered about how gene-transfer protocols might measure against these axes. Concerning the first, notwithstanding gene-transfer protocols aimed at dental caries or

**** Elsewhere, the Nuremberg Code posits "humanitarian importance" as the justification of risk.

erectile dysfunction (both of which are medically manageable), most current gene-transfer protocols target lethal and treatment-refractory diseases, and will thus tend to score high in terms of severity of unmet health need.

The temporal axis will tend to disfavor first-in-human trials of interventions like gene transfer, because such complex and novel strategies translate very slowly. The typical lag between phase 1 testing and licensure of a new drug is nine years. In the realm of biotechnology product development, a 20 percent annual discounting rate is generally applied. In addition, approximately 20 percent of biotechnology drugs that complete phase 1 studies are eventually licensed. Applying such discounting and risk figures to phase 1 gene-transfer studies greatly diminishes their value – at least as compared with trials in which launch into clinical application is immediate upon completion.[59]

The third consideration would in certain respects inflate gene-transfer trial value: many phase 1 trials in gene transfer benefit broad patient classes, because they produce knowledge than can be applied to many diseases. Thus, gene-transfer studies involving the various primary immunodeficiencies have proven informative for the development of strategies against various hematological disorders, like beta-thalassemia, hemophilia, or sickle cell disease. Arguably, the information gained from these studies is applicable to many non-hematological disorders as well.

Working against the dimension of population, however, are the questions of cost and access. In certain circumstances, gene-transfer technologies promise substantial cost savings to healthcare systems when compared with current standard of care. Gaucher's disease provides an example: current standard of care involves weekly or biweekly injections of the enzyme replacement product Cerezyme. Annual costs for the typical Gaucher's patient are about $200 000.[60] Gene transfer to stem cell tissues could, in principle, permanently obviate enzyme replacement. But a number of factors are likely to present significant hurdles for actualizing cost savings in gene transfer. First, these interventions often concentrate large expenditures at a single time-point, whereas chronic interventions like enzyme replacement spread cost throughout the life of a patient. Payers often resist large "one-off" costs, in part because of the possibility that other payers might accrue cost savings or that a patient will die before savings are realized. Second, in many instances, gene-transfer interventions will be categorized as orphan drugs. Orphan drug products tend to be exorbitantly expensive, largely because orphan drug laws extend 7 to 10 years of marketing exclusivity to manufacturers in order to encourage product development for rare disorders.[61] A third factor likely to affect cost and access is intellectual property. Gene-transfer technologies often embody multiple components; any single strategy will therefore involve a web of intellectual property claims. The costs for negotiating licenses from different patent holders,[62] as well as royalty stacking (that is, manufacturers paying royalties to different parties) may significantly increase the costs of many gene-transfer interventions.

Because gene transfer has not yet been commercialized, discussion about product pricing is largely speculative. Nevertheless, one can look to similar products to develop some sense of the challenge. One similar technology that is already available is bone-marrow transplantation. Encouragingly, the cost of bone-marrow transplantation has declined dramatically over several decades, largely because new techniques have improved the procedure's safety. Nevertheless, one set of commentators called this "one of the single most costly health care expenditures," estimating a cost between $41 000 and $85 000 per procedure, excluding the costs of hospitalization and complications.[63] Enzyme-replacement therapies provide another harbinger of how gene-transfer products might be priced. Take hemophilia: according to the National Hemophilia Foundation, the average annual cost of clotting factor replacement therapy for a typical patient ranges between $60 000 and $150 000.[64] One might expect that, because clotting factors have been on the market for a while and several companies produce clotting-factor products, the price of controlling hemophilia would have declined with time. According to one analysis, however, the contrary has occurred: over a nine-year period, expenditures on clotting factors at three major hemophilia treatment centers increased by an average of 17 percent a year. The inflationary trend related in large part to refinements in product that led to increased pricing.

Applying the humanitarian model

The humanitarian model of value greatly complicates the prospective assessment of study value. Assuming that the three dimensions of humanitarian value are exhaustive (they probably aren't), here are some of the imponderables that are presented: what constitutes a reasonable threshold for value along each of the three axes? Can some studies have a "negative" value along one of the dimensions – say, a medical enhancement technology that might undermine moral order by increasing social inequalities, or trials involving interventions that, if broadly applied, would bankrupt the healthcare system? How do the three axes relate to each other in the final appraisal of value: for instance, does immediate applicability to large populations compensate sufficiently for frivolity such that a risky wrinkle-cream study becomes permissible? A particular concern I have is how availability heuristics might bias value assessment in practice. For example, it is easy to imagine that the urgency of an unmet health need could overwhelm the population dimension during review, because relief of suffering for disease victims is easier to visualize than a person's inability to access a therapy because of cost. At any rate, how are IRBs, which generally lack broad policy, demographic, and economic expertise, to apply judgments about cost and access? And should IRBs even be in the business of nixing protocols because interventions are likely to be prohibitively costly?

I raise these questions not because I question the project of extending humanitarian criteria to translational trials, but rather, to point out the bewildering complexity of doing so. Of course, the same complexity confronts phase 3 studies: randomized controlled trials deemed valuable in terms of their design features might still be unacceptable ethically because they do not promise sufficient humanitarian benefits. What makes the humanitarian framework distinctively challenging for translational trials is the centerpiece of this book: the ethical confrontation with uncertainty. Of the three dimensions of humanitarian value described above, only one, the temporal, can be predicted with reasonable confidence. But which populations will ultimately benefit from a novel intervention study, whether the intervention will be affordable by the time it reaches later-stage studies, and the range of unmet health needs towards which the intervention (or variants thereof) will be applied is anybody's guess.

Nevertheless, several parameters can be offered that, because they reconcile humanitarian and translational models, provide a unified approach for appraising value. First, humanitarian considerations will generally favor greater attentiveness to a translational model of value because when translational studies are designed to yield reciprocal, iterative, and collateral knowledge, they widen the range of health needs and populations served by a study.

Second, though the applications towards which knowledge gathered in translational trials cannot always be known in advance, it is possible to specify the practical significance – both primary and ancillary – of a research undertaking. Thus, an ADA-SCID study is formally aimed at developing a treatment; secondary and tertiary objectives might include developing strategies for treating other hematological disorders and validating a technique that might

Table 6.1 Considerations for applying and reviewing translational trials

Design: negative findings should be amenable to meaningful interpretation
(1) Solid preclinical evidence (e.g. preclinical support for activity and mechanism)
(2) Flexible intervention (e.g. intervention can be modified if needed)
(3) Explicit description (e.g. composition of agents and delivery methods is well characterized)
(4) Target effects measured (e.g. correlative studies performed)
(5) Contingency plan (e.g. researchers have a plan in the event of "negative" findings)
Consent: study described as quasi-scientific investigation
(1) Rigorous consent process emphasizing nontherapeutic nature of study (especially correlative component)

be applied to a wide variety of disorders. Notwithstanding the considerable uncertainty surrounding the ultimate application of an intervention tested in a translational trial, these goals at least provide guideposts for a discussion of humanitarian value.

Third, for reasons of administration and policy, ethics review bodies should be very cautious about second-guessing value assessments occurring downstream in the research process *in circumstances where innovation incentives and practices can be said to be reasonably aligned with the needs of host populations.* That is, at the point where gene-transfer interventions are entering phase 1 studies, the public will have made significant investments in the basic and applied research necessary to bring the agent to trials. No doubt that research policy in industrialized democracies is deeply flawed, with certain health needs left under-researched, other health areas receiving disproportionate attention because of political access, research heavily skewed towards commercial ends, and little consideration of healthcare sustainability. Nevertheless, the most appropriate forum for serious ethical deliberation about humanitarian significance is at the point of policy (by which I mean funding and the development of research incentives). After that, IRBs should have clear and compelling reasons to hold up a first-in-human study on broad humanitarian grounds. These might include circumstances where there is reason to question whether innovation policy and practice are mismatched with population health needs (this will often raise questions of fairness; see Chapter 8), or about applications whose humanitarian potential is trivial when measured against a study's risks (for example, a first-in-human study aimed at developing an enhancement application). It nevertheless seems not inappropriate for IRBs to exhort researchers and sponsors to tackle issues they consider more pressing from a humanitarian angle.

Conclusion

First-in-human trials have serious constraints in terms of value: they are statistically underpowered, riddled with bias, and because of concerns about risk, tend to enroll populations that are least likely to show the desired biological responses. Nevertheless, the profound uncertainty confronting investigators who wish to test a novel intervention represents an opportunity: when uncertainty is great enough, even small amounts of information clarify the issues and allow the formulation of clear, testable hypotheses.

This chapter proposes a model of value that better respects the volunteer because it improves the probability that translational studies will be informative. Implications for design, and consent procedures are summarized in Table 6.1. The translational model proposed here better captures the reality of how information flows in translational research. According to this model, studies should be planned and designed to be informative in the (likely) event interventions do not manifest intended biological effects. Specifically, studies should provide a starting point for trouble-shooting applications (or preclinical models) that otherwise showed promise in the laboratory, and for informing less immediately related lines of inquiry.

References

1. Jansen B, Wacheck V, Heere-Ress E, Schlagbauer-Wadl H, Hoeller C, Lucas T, *et al.* Chemosensitisation of malignant melanoma by BCL2 antisense therapy. *Lancet* 2000; **356**(9243): 1728–33.
2. Piantadosi S. *Clinical Trials: A Methodologic Perspective*, 2nd edn. Hoboken, NJ, Wiley Interscience, 2005; 225.
3. O'Brien SM, Cunningham CC, Golenkov AK, Turkina AG, Novick SC, Rai KR. Phase I to II multicenter study of oblimersen sodium, a Bcl-2 antisense oligonucleotide, in patients with advanced chronic lymphocytic leukemia. *J Clin Oncol* 2005; **23**(30): 7697–702.
4. Bedikian AY, Millward M, Pehamberger H, Conry R, Gore M, Trefzer U, *et al.* Bcl-2 antisense (oblimersen sodium) plus dacarbazine in patients with advanced melanoma: the Oblimersen Melanoma Study Group. *J Clin Oncol* 2006; **24**(29): 4738–45.
5. Nuremberg Code. In: *Trials of War Criminals before the Nuremberg Military Tribunals under Control Council Law No. 10, Vol. 2.* Washington, DC, US Government Printing Office, 1949; 181–2.
6. Evans JH. *Playing God?: Human Genetic Engineering and the Rationalization of Public Bioethical Debate*. Chicago, University of Chicago Press, 2002.
7. Eggermont AM. Reaching first base in the treatment of metastatic melanoma. *J Clin Oncol* 2006; **24**(29): 4673–4.
8. Fergusson D, Doucette S, Glass KC, Shapiro S, Healy D, Hebert P, *et al.* Association between suicide attempts and selective serotonin reuptake inhibitors: systematic review of randomised controlled trials. *BMJ* 2005; **330**(7488): 396.
9. Avorn J. *Powerful Medicines: The Benefits, Risks, and Costs of Prescription Drugs.* Chapter 7. New York, Vintage Books, 2005.
10. Fleming TR, DeMets DL. Surrogate end points in clinical trials: are we being misled? *Ann Intern Med* 1996; **125**(7): 605–13.
11. Schwartz D, Lellouch J. Explanatory and pragmatic attitudes in therapeutical trials. *J Chronic Dis* 1967; **20**(8): 637–48.
12. Tunis SR, Stryer DB, Clancy CM. Practical clinical trials: increasing the value of clinical research for decision making in clinical and health policy. *JAMA* 2003; **290**(12): 1624–32.
13. Lee PY, Alexander KP, Hammill BG, Pasquali SK, Peterson ED. Representation of elderly persons and women in published randomized trials of acute coronary syndromes. *JAMA* 2001; **286**(6): 708–13.
14. Halpern SD, Karlawish JH, Berlin JA. The continuing unethical conduct of underpowered clinical trials. *JAMA* 2002; **288**(3): 358–62.
15. http://edocket.access.gpo.gov/cfr_2005/aprqtr/pdf/21cfr312.21.pdf.
16. Kay MA, Manno CS, Ragni MV, Larson PJ, Couto LB, McClelland A, *et al.* Evidence for gene transfer and expression of factor IX in haemophilia B patients treated with an AAV vector. *Nat Genet* 2000; **24**(3): 257–61.
17. Manno CS, Pierce GF, Arruda VR, Glader B, Ragni M, Rasko JJ, *et al.* Successful transduction of liver in hemophilia by AAV-Factor IX and limitations imposed by the host immune response. *Nat Med* 2006; **12**(3): 342–7.
18. Hasbrouck NC, High KA. AAV-mediated gene transfer for the treatment of hemophilia B: problems and prospects. *Gene Ther* 2008; **15**(11): 870–5.
19. Orkin SH, Motulsky AG. Report and Recommendations of the Panel to Assess the NIH Investment in Research on Gene Therapy. US National Institutes of Health, Office of Biotechnology Activities, 1995.
20. Grobstein C, Flower M. Gene therapy: proceed with caution. *Hastings Cent Rep* 1984; **14**(2): 13–17.
21. Albelda SM, Sterman DH. TNFerade to the rescue? Guidelines for evaluating phase I cancer gene transfer trials. *J Clin Oncol* 2004; **22**(4): 577–9.
22. DeYoung MB, Dichek DA. Gene therapy for restenosis: are we ready? *Circ Res* 1998; **82**(3): 306–13.

23. Wu Z, Sun J, Zhang T, Yin C, Yin F, Van Dyke T, *et al.* Optimization of self-complementary AAV vectors for liver-directed expression results in sustained correction of hemophilia B at low vector dose. *Mol Ther* 2008; **16**(2): 280–9.

24. McCaffrey AP, Fawcett P, Nakai H, McCaffrey RL, Ehrhardt A, Pham TT, *et al.* The host response to adenovirus, helper-dependent adenovirus, and adeno-associated virus in mouse liver. *Mol Ther* 2008; **16**(5): 931–41.

25. Jiang H, Couto LB, Patarroyo-White S, Liu T, Nagy D, Vargas JA, *et al.* Effects of transient immunosuppression on adenoassociated, virus-mediated, liver-directed gene transfer in rhesus macaques and implications for human gene therapy. *Blood* 2006; **108**(10): 3321–8.

26. Zaiss AK, Cotter MJ, White LR, Clark SA, Wong NC, Holers VM, *et al.* Complement is an essential component of the immune response to adeno-associated virus vectors. *J Virol* 2008; **82**(6): 2727–40.

27. Piantadosi S. Translational clinical trials: an entropy-based approach to sample size. *Clin Trials* 2005; **2**(2): 182–92.

28. US Food and Drug Administration. *Innovation or Stagnation? Challenge and Opportunity on the Critical Path to New Medical Products.* US Department of Health and Human Services, 2004.

29. Kimmelman J. Ethics at phase 0: clarifying the issues. *J Law Med Ethics* 2007; **35**(4): 727–33.

30. Christian M, Shoemaker D. *The Investigator's Handbook: A Manual for Participants in Clinical Trials of Investigational Agents Sponsored by DCTD, NCI.* National Cancer Institute, 2002.

31. National Cancer Institute. *Clinical Trials: Questions and Answers. NCI Fact Sheet.* US National Institutes of Health. www.cancer.gov/cancertopics/factsheet/Information/clinical-trials; 2006.

32. American Cancer Society. *Clinical Trials: What You Need to Know.* www.cancer.org/docroot/ETO/content/ETO_6_3_Clinical_Trials_-_Patient_Participation.asp#C4; 2008.

33. Cancer Research UK. *About Cancer Research: Phase 1 Trials.* http://info.cancerresearchuk.org/cancerandresearch/aboutcancerresearch/differentareasofresearch/clinical/phase1trials/; 2004.

34. Horstmann E, McCabe MS, Grochow L, Yamamoto S, Rubinstein L, Budd T, *et al.* Risks and benefits of phase 1 oncology trials, 1991 through 2002. *N Engl J Med* 2005; **352**(9): 895–904.

35. *Nature Medicine.* Press Release: Gene therapy combined with chemotherapy effective against cancer. www.nature.com/nm/press_release/nm0800.html; August 2000.

36. Varghese S, Rabkin SD, Nielsen PG, Wang W, Martuza RL. Systemic oncolytic herpes virus therapy of poorly immunogenic prostate cancer metastatic to lung. *Clin Cancer Res* 2006; **12**(9): 2919–27.

37. Miller ML. *Successful Phase I gene therapy trial with PD patients.* www.hdlighthouse.org/.

38. Gershon ES. Making progress: does clinical research lead to breakthroughs in basic biomedical sciences? *Perspect Biol Med* 1998; **42**(1): 95–102.

39. Stokes DA. *Pasteur's Quadrant: Basic Science and Technological Innovation.* Washington, DC, Brookings Institution Press, 1997.

40. Gelijns AC, Thier SO. Medical innovation and institutional interdependence: rethinking university–industry connections. *JAMA* 2002; **287**(1): 72–7.

41. Li JY, Englund E, Holton JL, Soulet D, Hagell P, Lees AJ, *et al.* Lewy bodies in grafted neurons in subjects with Parkinson's disease suggest host-to-graft disease propagation. *Nat Med* 2008; **14**(5): 501–3.

42. Kordower JH, Chu Y, Hauser RA, Freeman TB, Olanow CW. Lewy body-like pathology in long-term embryonic nigral transplants in Parkinson's disease. *Nat Med* 2008; **14**(5): 504–6.

43. Mendez I, Vinuela A, Astradsson A, Mukhida K, Hallett P, Robertson H, *et al.* Dopamine neurons implanted into people with Parkinson's disease survive without pathology for 14 years. *Nat Med* 2008; **14**(5): 507–9.

44. Braak H, Del Tredici K. Assessing fetal nerve cell grafts in Parkinson's disease. *Nat Med* 2008; **14**(5): 483–5.

45. Moed HF. *Citation Analysis in Research Evaluation.* Dordecht, The Netherlands, Springer, 2005.

46. Soares HP, Kumar A, Daniels S, Swann S, Cantor A, Hozo I, *et al.* Evaluation of new treatments in radiation oncology: are they better than standard treatments? *JAMA* 2005; **293**(8): 970–8.
47. Kumar A, Soares H, Wells R, Clarke M, Hozo I, Bleyer A, *et al.* Are experimental treatments for cancer in children superior to established treatments? Observational study of randomised controlled trials by the Children's Oncology Group. *BMJ* 2005; **331**(7528): 1295.
48. Joffe S, Harrington DP, George SL, Emanuel EJ, Budzinski LA, Weeks JC. Satisfaction of the uncertainty principle in cancer clinical trials: retrospective cohort analysis. *BMJ* 2004; **328**(7454): 1463.
49. Kola I, Landis J. Can the pharmaceutical industry reduce attrition rates? *Nat Rev Drug Discov* 2004; **3**(8): 711–15.
50. Camacho LH, Bacik J, Cheung A, Spriggs DR. Presentation and subsequent publication rates of phase I oncology clinical trials. *Cancer* 2005; **104**(7): 1497–504.
51. Goulart BHL, Roberts TG, Clark JW. Utility and costs of surrogate endpoints (SEs) and biomarkers in phase I oncology trials. *J Clin Oncol (Meeting Abstracts)* 2004; **22**: 6012.
52. Malakoff D. Pentagon agency thrives on in-your-face science. *Science* 1999; **285**(5433): 1476–9.
53. Committee on Prospering in the Global Economy of the 21st Century, National Academy of Sciences, National Academy of Engineering, Instute of Medicine. *Rising Above the Gathering Storm: Energizing and Employing America for a Brighter Economic Future*. Washington, DC, The National Academies Press, 2007.
54. Mervis J. US science policy. Risky business. *Science* 2004; **306**(5694): 220–1.
55. Emanuel EJ, Wendler D, Grady C. What makes clinical research ethical? *JAMA* 2000; **283**(20): 2701–11.
56. Grady C. Science in the service of healing. *Hastings Cent Rep* 1998; **28**(6): 34–8.
57. Freedman B. Placebo-controlled trials and the logic of clinical purpose. *IRB* 1990; **12**(6): 1–6.
58. Casarett DJ, Karlawish JH, Moreno JD. A taxonomy of value in clinical research. *IRB* 2002; **24**(6): 1–6.
59. Stewart JJ, Allison PN, Johnson RS. Putting a price on biotechnology. *Nat Biotechnol* 2001; **19**(9): 813–17.
60. Anand G. A biotech drug extends a life, but at what price? *The Wall Street Journal* 2005 November 16.
61. Drummond MF, Wilson DA, Kanavos P, Ubel P, Rovira J. Assessing the economic challenges posed by orphan drugs. *Int J Technol Assess Health Care* 2007; **23**(1): 36–42.
62. Heller MA, Eisenberg RS. Can patents deter innovation? The anticommons in biomedical research. *Science* 1998; **280**(5364): 698–701.
63. Westerman IL, Bennett CL. A review of the costs, cost-effectiveness and third-party charges of bone marrow transplantation. *Stem Cells* 1996; **14**(3): 312–19.
64. National Hemophilia Foundation. Financial and Insurance Issues. www.hemophilia.org/NHFWeb/MainPgs/MainNHF.aspx?menuid=34&contentid=24; 2006.

Chapter 7

The chasm: the ethics of initiating first-in-human clinical trials

Introduction

In June of 2001, a team of gene-transfer researchers led by Matthew During and Michael Kaplitt presented the RAC with a clever new strategy against Parkinson's disease. The second most common neurodegenerative disorder in North America, the cardinal symptoms of Parkinson's disease – tremor, rigidity, and inability to initiate movement – are caused in part by the excessive firing of a structure deep inside the brain called the subthalamic nucleus. The researchers proposed to genetically modify this structure with a gene encoding the inhibitory neurotransmitter, glutamic acid decarboxylase (GAD). With the subthalamic nucleus churning out GAD, the investigators postulated that nearby brain structures would quiet down, and Parkinsonian symptoms would abate.

The protocol required that a surgeon insert a needle through the volunteer's brain and inject small quantities of vector. But you didn't need to be a brain surgeon to recognize the peril. This would be the first administration of AAV vectors to the brains of non-terminal patients, and the investigators could not rule out the possibility that the vector might trigger an autoimmune reaction. Nor, according to members of the RAC, had the investigators decisively established the efficacy of their intervention in non-human primates.[1] Though the RAC never formally advised against initiating the study, controversy trailed the team out of Bethesda. When the study was initiated in 2003, several leading Parkinson's disease researchers denounced it as "a crazy experiment" and "terra incognita."[2] Another accused the lead researcher of "raising hopes in people with minimal evidence of benefits."[3]

Were the members of the RAC and various critics too restrictive, bullying a group of enterprising researchers? Or was the RAC too permissive in its gentle requests for a "discussion … of what animal models would be relevant for this study?" What criteria should investigators and committees use when considering a translational trial? For that matter, what factors do investigators and committees consider in making such decisions, and are these appropriate?

Surprisingly, in over thirty years of research ethics scholarship, the field has yet to address these questions in a systematic fashion (see Box 7.1).[4] This is a puzzling oversight, and one that is increasingly problematic in the light of a series of new initiatives (described in Chapter 2) spurring first-in-human research.

What follows is an attempt to map the major ethical questions. I propose a framework for deciding when to initiate testing, and enumerate factors that should be considered by investigators and ethics committees. I close with some thoughts about research avenues that might prove helpful in refining this framework.

AAV-GAD on the brain

After Kaplitt and During pitched their protocol to the RAC, they offered several defenses. They underscored their procedure's safety: in their studies in monkeys and rats, they had never once observed behaviors or deficits that would raise concern about toxicity. Even if safety problems arose, the team had a "rescue strategy" waiting in the wings: researchers would lesion the brain structure receiving vector. Since subthalamotomies are sometimes performed to relieve symptoms of Parkinson's disease, the "rescue" would leave volunteers in a state not much different than other Parkinson's patients in later stages of their disease. They further argued that the theory supporting their strategy was sound. Brain surgeons frequently implant electrodes in the subthalamic nucleus of Parkinson's patients in order to dampen this brain structure's activity (this therapy is known as deep-brain stimulation). Rather than implanting an electrode, the researchers proposed to use a vector to do the same thing. Kaplitt and During further contended that it was unclear whether primate efficacy data had any relevance for predicting human efficacy. After all, primate models of Parkinson's disease are notoriously problematic. Human disease is typically chronic and degenerative; primate models are based on acute injuries and are not necessarily degenerative. Human disease is typically spontaneous and occurs with aging. Primate models involve inducing Parkinsonian symptoms by delivering a neurotoxin, 1-methyl-4-phenyl-1,2,3,6-tetrahydropyridine (MPTP), to the brain; monkeys used in these studies are typically younger than their human counterparts. Given these deficiencies, why bother with animal evidence at all?

Critics of the protocol, like RAC member Theodore Friedmann, responded with several arguments. He rejected the team's claims about safety, arguing that brain chemistry is delicately poised such that any disruption could "wreak havoc" in a way that would not be detected in rodent studies. Friedmann also argued that the protocol involved a "unique" approach for intervening against Parkinson's disease. Such novelty created "all the more reason to prove it to the hilt … The natural stopping point before you go into the clinic is to pick absolutely the most relevant animal model that exists," he argued. For During and Kaplitt, this would have involved demonstrating efficacy in MPTP-treated monkeys.[5]

The debate between Kaplitt and his critics centered over four major fault lines. First, is preclinical evidence of efficacy (in what follows, I will call preclinical results that indicate possible human efficacy as "supporting evidence") really necessary? One might argue that all that really matters is safety. This position might strike many readers as implausible. Nevertheless, it is worth noting that when the Food and Drug Administration evaluates proposals to initiate phase 1 studies, projected efficacy is not a central consideration.[6] Why, then, require any animal evidence of efficacy?

Second, is supporting evidence dispensable when there are credible grounds to doubt the probative value of preclinical models? While Kaplitt seemed to answer this question in the affirmative, Friedmann seemed to suggest that, in such circumstances, efficacy should nevertheless be established in the best available animal model.

Third, does the novelty of an approach underwrite an ethical obligation to exhaust research in animals before pursuing studies in humans? Here, Friedmann took a strong position. Whether Kaplitt and others would disagree is unclear; they might instead argue that their approach is not all that novel, since parallel interventions like deep-brain stimulation are used routinely in clinical practice.

Fourth, should the stringency of supporting evidence increase with risk? The above discussion hardly exhausts the variables that might be considered in deciding how much and what type of supporting evidence should be demanded. Might, for example, the possibility of therapeutic gain from participation justify a relaxed evidentiary standard?

Box 7.1 Extending existing ethical frameworks for first-in-human trials

How do investigators and ethicists currently approach the ethics of launching first-in-human trials? And what are some of the limitations of these policies?

Initiation of a trial will depend on a judgment on whether expected benefits – direct or scientific – are favorably balanced with risks. Thus, for example, Jeremy Sugarman, writing on initiating in utero gene-transfer studies, commented that "the experiment itself must be designed in such a way as to produce useful results" and that "accurate forecasting about safety in proposed human experimentation is essential ..." Sensible though these statements are, what counts as a "useful result," and what are the conditions most conducive to their generation? What counts as "accurate forecasting?" And what factors should investigators, policy-makers, ethicists, and IRBs consider in addressing such questions?

One option would be to apply an off-the shelf concept used elsewhere in medical ethics for deciding clinical trial initiation: clinical equipoise. As described by the concept's originator, Benjamin Freedman, clinical equipoise refers to a lack of "consensus within the expert clinical community about the comparative merits of the alternatives to be tested."[7] Various commentators have generalized the concept beyond controlled clinical trials with the suggestion that a study is in equipoise provided that no volunteer will receive an intervention unless a reasonable proportion of the expert medical community regards it as equal to or possibly superior to the standard of care.

Equipoise has several attractive features – not the least of which is that it provides a structure for deciding the risk–benefit and value acceptability for a proposed intervention. However, I am not convinced that it is applicable to first-in-human studies. Some of the reasons are described in Chapter 5.[8] Specifically, translational trials begin with what one commentator calls "complete ignorance." That is, whereas clinicians can form a credible opinion about a drug by the time it has reached phase 3 testing, the clinician's uncertainty is, in a sense, unbounded for novel compounds entering phase 1 testing.[9] The "robust epistemic threshold"[10] of equipoise, coupled with the paramount obligation to "do no harm," requires that competent medical practitioners assume that novel interventions (that is, interventions for which there is no clinical experience) are useless and possibly even harmful unless there is clinical evidence to the contrary. A second problem with equipoise concerns its value component: according to Freedman and others, ethical trials should be capable of perturbing clinical equipoise upon completion. Yet it is difficult to see precisely how translational trials, structured as they are around bounding uncertainty, could ever perturb equipoise.

An alternative approach might be to develop formalized criteria: investigators might articulate specific conditions for initiating trials. For example, in the immediate aftermath of the Cline affair in 1980, W. French Anderson and ethicist Joseph Fletcher published an opinion piece that proposed three criteria: transgenes should "be put into the proper target cells and should remain there; they should be expressed appropriately; and they should not harm transduced cells."[11] Whether such conditions are sufficient can be debated. Clearly the Gene Therapy Working Group within the Recombinant DNA Advisory Committee did not think they were, and history would appear to have vindicated them. Other field-specific guidelines have been established for xenotransplantation[12] and neuroregeneration in Alzheimer's disease and Parkinson's disease clinical trials.[13] These offer the advantages of integrating expertise, establishing research goals, providing clear progress benchmarks, and rendering decision-making about trial initiation more transparent. However, formalized criteria must be reinvented for each innovation and are quickly outmoded in cutting-edge research areas. Second, criteria are often vague – what exactly does it mean, for example, to say that transgenes are "expressed appropriately?" Last, absent a clear ethical framework, the research community would lack any basis for judging the ethical adequacy of criteria. Indeed, the criteria provided by Anderson and Fletcher proved unpersuasive for many RAC reviewers (see Chapter 9). For these reasons,

Box 7.1 (*Cont.*)

formalized criteria alone do not provide a satisfying answer to the question of when trials should be initiated.

A third approach is procedural. Trial initiation might be decided formally through a tribunal structure like that of the RAC.[14] Alternatively, initiation might be decided more informally through the emergence of professional consensus. For example, throughout the 1950s and 1960s, the polio vaccine research community censured peers who pushed their vaccine candidates too aggressively into human studies.[15] Similar dynamics regulated the testing of various transplant surgeries.

Like formalized criteria, procedural approaches marshal expertise and have much to commend them. However, they have several shortcomings. First, they vary widely among fields: formal structures like the RAC do not exist for other translational platforms and would need to be created each time a new intervention is proposed for clinical testing. Second, informal strategies, which worked reasonably well for certain types of organ transplantation[16] (but not others)* might be more difficult to sustain where commercial or competitive pressures undermine professional cohesion. Third, as with the formalized approach, without an ethical framework, procedural mechanisms cannot be assessed for their ethical justifiability. One can imagine that some procedural structures might approach a given protocol restrictively, while others might approach the same protocol permissively. Indeed, procedural mechanisms tend to work best when they can draw on policies that structure their reasoning (for typical research protocols, these might include the Declaration of Helsinki, the Belmont Report, or institutional policies).

The ethical justification of preclinical evidence

The first question asked above was whether preclinical evidence of efficacy is necessary. Why aren't hunches and predictions based on mechanism and basic biology sufficient? Why must researchers actually present empirical data – typically a preclinical study comparing active intervention against a series of controls – to support their human studies?

One way of answering this cluster of questions would be to look at various well-established practices and institutions for guidance, and ask whether opting to forgo the collection of supporting data coheres with these. On the one hand, Chapter 5 examined drug-regulation policy, and found that US FDA standards for supporting evidence are not focused on efficacy. On the other hand, many ethical codes mandate supporting evidence. The Nuremberg Code states "the [human] experiment should be … based on the results of animal experimentation and knowledge of the natural history of the disease … that *the anticipated results will justify the performance of the experiment*."[18] Similarly, the Council for International Organizations of Medical Sciences (CIOMS) states "clinical testing must be preceded by adequate laboratory or animal experimentation to demonstrate *a reasonable probability of success* without undue risk."[19] The italicized portions of both statements (both of which are added) emphasize that the ethical basis for collecting preclinical data is not merely to demonstrate safety, but also to establish the plausibility of the hypothesis tested in human beings.

Another approach to answering this cluster of questions is to explore the ethical basis for requiring supporting data. I argued (also in Chapter 5) that the ethical assessment for novel agent translational trials – perhaps for non-novel ones as well – requires that risks be justified by the biomedical value of the knowledge gained rather than the prospect of direct medical benefit for the volunteer. In Chapter 4, I argued that the ethical appraisal of risk should begin

* This specifically refers to Denton Cooley's attempt at transplanting sheep's hearts into human patients.[17]

with "upper-bound" estimates of the probability of harm. A consequence of combining these two arguments is that first-in-human trials should have a high probability of producing useful biomedical knowledge.

My thesis is that human studies founded on a solid preclinical evidence base are generally more informative – and thus more likely to be valuable to the broader scientific community – than studies launched on "hunches" or postulation from mechanism.† The thesis rests on the following reasoning.

One of the most dreaded questions scientists receive when they present experimental results to other scientists is "why didn't you run a control for X?" The query signals disagreement with how an experimenter interprets her results. Imagine a dialog between a researcher and her "control-seeking" interlocutor. The researcher asserts that, when she injects vector X into tumor-bearing mice, their cancer goes away. "How do we know that the animals didn't just mount a non-specific immune response against the cancer?" asks the skeptic. "Well," says the researcher, "I ran a control with untreated animals, and these ones died." "Fine" says the skeptic. "But the vector itself might have triggered some immune response against the cancer. How do we know your transgene is doing anything?" "Because my control animals actually received vector expressing a benign transgene. Now do you believe me?" "No I don't. You don't even know whether the transgenes in either arm were expressed." "Yes I do. I performed immunostaining and RT-PCR on samples from the tumor tissues in both groups. The transgenes were expressed in both active and control arms of my study." "Hmmm," says the skeptic. Thinking 'this researcher strikes me as a shady character; these results are too good to be true,' the skeptic says "to be on the safe side, I would like to see the same results in someone else's hands." And so on.

Scientists and clinicians tend to form stronger beliefs about a hypothesis when credible competing explanations can be virtually refuted. How does this relate to the launch of preclinical testing? Human studies that are launched on strong beliefs will tend to be more informative than those that aren't, because either they will unexpectedly confirm those beliefs (put in statistical language, they will refute the null hypothesis that expected effects will not occur), or they will falsify one or multiple strong beliefs that underwrite the human experiment. That is, a translational human trial can have two stereotypical outcomes: either the agent reaches its target and induces desired biological effects without causing toxicity (the "positive result") or the agent fails either because it doesn't cause the intended biological effects or because it is unsafe (the "negative result"). Experience tells us that "positive results" for translational trials are uncommon. Recall, for example, that about 5 percent of new chemical entities entering phase 1 testing against cancer will survive through to commercialization; for drugs that operate by novel mechanisms, the rate is likely to be lower.[20] Another study that drives this point home found that only 37 percent of preclinical animal studies published in top-tier biomedical journals were replicated in human randomized trials; fewer than a third of these drugs were ever licensed for human use.[22] Clearly, then, a "positive result" is unexpected and highly informative.

† I note here that I am not aware of any empirical studies documenting that first-in-human trials produce more valuable biomedical knowledge when they are based on extensive and high-quality supporting evidence. Nevertheless, some commentators have asserted a relationship between the quality and extent of preclinical testing and likelihood that a new compound will show efficacy in subsequent clinical trials. For examples see Kola & Landis[20] and Pangalos *et al.*[21]

The negative results are not particularly informative when there are obvious explanations for them. Returning to the dialog above, imagine that the zealous preclinical investigator did not use a conservative method to assure tumor take and did not perform blinded outcome assessment. When her clinical colleagues get a negative result in a phase 1 trial, the skeptic is perfectly justified in viewing the study as a waste of time and resources. On the other hand, if the preclinical researcher has performed extensive and rigorous testing, the negative result contains information about physiological and/or pathological differences between the model animal and the modeled patient. This might lead to new lines of inquiry aimed at understanding the nature of the difference. Or, knowing about the difference between the human and animal, the researchers now have a better idea of how to interpret the results of preclinical experiments. Or, if the investigators can discover the source of that difference, they can then modify their strategy to overcome these differences so that the animal model becomes more predictive.[‡] Thus, to quote the Orkin–Motulsky report, "[s]tudying the differences between human diseases and animal model phenotypes may provide insights into disease pathogenesis that may, in turn, be exploited either by gene therapy or pharmacological approaches."[23]

From this analysis, I draw two conclusions. First, hunches and guesses, however well-informed by mechanistic knowledge, do not provide sufficient justification for launching novel intervention studies. Second, translational trials tend to be more valuable when they are buttressed by strong supporting evidence, because either they unexpectedly confirm an effect, or they provide some indication about diagnosing a negative result.

Scaling evidentiary requirements according to availability of animal models

Should the quantity and quality of evidence be tightened in some instances, or relaxed in others? One variable suggested in the debate between Kaplitt and Friedmann is the quality of animal models. Some diseases have animal models that recapitulate the essential features of human disease. Hemophilia is one example: several research centers maintain colonies of dogs that manifest pathologies that are nearly identical to those of human beings. On the other hand, neurodegenerative disorders like Parkinson's disease are very difficult to model effectively, as there are no known spontaneous disease models. Policy-makers might demand that researchers make use of faithful models before initiating human studies. On the other hand, policy-makers might allow a phase 1 trial to proceed despite the fact that supporting data are far less predictive. In essence, then, one might scale evidence according to "natural barriers" presented by the limitations of existing animal models.

This approach is intuitive enough, but it brings up several problems. The first centers on its implications for the risk–benefit balance for a phase 1 study. Scaling evidence according to model availability might not optimize the probability of gathering useful knowledge. Consider a scenario in which investigators are contemplating initiating trials against two human diseases, D_a and D_b, that have identical mortality and morbidity. Let's also assume that interventions, I_a and I_b, carry identical risks. Let's further assume that D_a researchers use

[‡] For a good example, consider some of the hemophilia gene-transfer trials described in Chapter 6 under the section "The translational model of value." These studies uncovered unexpected differences between human and animal immune responses and vector biodistribution. As a result, scientists have a better basis for interpreting preclinical studies involving similar interventions.

an animal model that recapitulates the major features of the human illness, while the standard model used in preclinical studies for D_b is notoriously unreliable. Lastly, let's imagine that in preclinical studies, investigators achieved equally striking reversals of disease using I_a and I_b.

For reasons outlined above, the study I_a is more likely than I_b to produce valuable knowledge regardless of whether the results are positive or negative. For I_b, a negative result is minimally informative and does not lead to any obvious new lines of inquiry. Given the premises with which we began, the risk:benefit ratio of the I_b phase 1 trial is inferior to that of I_a. This does not necessarily make the phase 1 trial of I_b unethical, because we might still consider trial B to have an *acceptable* risk–benefit balance. Nevertheless, if the analysis above holds, it suggests that scaling evidence requirements according to the quality of existing model organisms diminishes the value of the knowledge gained from a translational study. In so doing, the ratio of predicted social benefit to risk diminishes.

A second reason why scaling evidence according to the reliability of models is unsatisfactory is that it makes both the limits of knowledge as well as the relationship between preclinical and clinical studies appear to be detemined by nature rather than human decision-making. For instance, instead of pursuing first-in-human studies in human volunteers, researchers might devote greater energy toward developing better animal models. Or, instead of testing a new drug in a phase 1 dose-escalation study, investigators might perform a pilot human study to determine whether the molecular responses seen in animals can be reproduced in human beings. For example, instead of organizing a study around the hypothesis that a therapeutic response can be induced in volunteers with cancer, investigators might perform smaller studies asking whether a vector can reach the desired target, and if so, whether there is any evidence of transgene expression and cellular response. Armed with confirmatory evidence, they might then proceed to a phase 1 study in which escalating doses of vector are delivered to cancer patients at a level necessary to trigger response.

As will be explored below, smaller-scale "exploratory" studies will often be the appropriate course of action where animal models are severely deficient. Nevertheless, there are exceptional circumstances where performing exploratory studies could expose volunteers to burdens and levels of risk that are roughly comparable to those in a phase 1 dose-escalation study. In these rare instances, an absence of adequate preclinical models might very well justify a relaxation of evidentiary standards for trial initiation.

Risk, illness, and evidentiary standards

Another factor that might drive standards of supporting evidence is risk. In discussions centering on Kaplitt's Parkinson's disease protocol, Friedmann asserted that the potential unintended consequences of tampering with delicate brain chemistry necessitate the highest possible level of supporting evidence.

That a protocol's increased risk obliges a higher standard of evidence should be uncontroversial. This is, after all, how most functional institutions approach risk. Banks making larger loans generally gather more intelligence on their investment. Physicians administering risky drugs or procedures will generally want to have greater confidence that their intervention has therapeutic value (though in practice, this is not always the case: consider the example of surgery!). As for drug regulation, the US FDA normally requires that pharmaceutical companies establish improved survival or symptom relief before licensing a new cancer treatment. However, when new drugs are relatively nontoxic, the agency is willing to tolerate greater uncertainty about clinical efficacy – at least for cancer drugs – by accepting the use of surrogate endpoint data.[24] Such policies and practices are consistent with scaling supporting evidentiary requirements with risk.

This position also finds support in the reasoning advanced in previous sections. I suggested that better supporting evidence generally enhances the scientific value of a protocol. If a protocol involves higher risk, maintaining an appropriate risk–benefit balance requires that scientific value be strengthened. It follows, then, that riskier protocols should be supported by stronger preclinical evidence.

What about the suggestion that illness severity provides a justification for lowering evidence standards (or, recast in a positive way, that when investigators propose risky studies in volunteers with moderate conditions, evidence standards should be tightened)? Parkinson's disease has an inexorably morbid and fatal course, while a disease like hemophilia can be managed effectively using coagulant-factor replacement therapy. On these grounds, should investigators produce stronger supporting evidence for hemophilia trials than for Parkinson's disease trials?

An affirmative answer turns out to have very complex implications. Based on the reasoning in previous sections, diminishing the evidentiary standards for initiating trials in treatment-refractory volunteers would seem to diminish the expected value of such studies. This, in turn, translates to a less favorable risk–knowledge balance for such studies. Many would rightly condemn a policy as regressive and unjust if it allowed investigators to expose terminally ill patients to greater risks in order to gain a unit increase in biomedical knowledge (recall that in previous chapters, I have emphasized that risks in such early-phase studies are justified by knowledge value rather than therapeutic benefits).

However, risks, harms, and burdens will vary according to a volunteer's disease status. The potential harms and burdens for a typical gene-transfer study illustrate this point. Put crudely, immediate death exacts less opportunity cost for the terminally ill volunteer than it would for someone expected to live for another 30 years. The terminal volunteer is also much less likely than the stable one to survive long enough to develop a slow-growing malignancy. On the other hand, pain from study interventions will not be experienced differently by terminal and stable volunteers. The time commitments and schedule disruption might actually be more burdensome for terminal volunteers, since time is exactly what they lack. All this makes for an incredibly messy analysis, and it seems next to impossible to offer any general statements to the effect that, when volunteers are sicker, lower evidentiary standards are (or are not) acceptable.

In summary, stronger evidence should generally be demanded where translational studies involve greater risk. Though in some circumstances, increased disease severity will translate into diminished risk, relaxing evidence standards where protocols enroll terminally ill volunteers has no clear justification.

The principle of translational distance: a common framework for value in first-in-human trials

In Chapter 4 I argued that studies involving the first systemic exposure of a human being to biologically active levels of a novel intervention present a high degree of uncertainty about safety and response. This uncertainty translates to (at the very least) a moderate degree of risk, and makes the study's value all the more speculative.

I propose that all first-in-human translational trials that involve novel agents should satisfy a common threshold of evidentiary support, much as all later-stage clinical trials must satisfy a common threshold of validity. Once this threshold has been met, investigators, policy-makers, and IRBs might ask for further supporting evidence depending on the study's degree of risk.

How might a common threshold of evidence be defined? Consider the nature of the preclinical experiment. When a researcher conducts a laboratory experiment, she uses observations collected in a controlled setting to project what will occur in the clinic. This requires that she maintain numerous assumptions about her preclinical experiment's similarity to a clinical circumstance.[25] Some preclinical tests (e.g. a hemophilia gene-transfer study) demand fewer or less radical assumptions because animal models faithfully recapitulate essential features of human pathology. Others, like cystic fibrosis or Parkinson's disease, will often demand greater or more extravagant assumptions. The process of projection is especially prone to error in situations like these. In considering the ethics of trial initiation, the dissimilarity between preclinical experiments and clinical trials might be likened to a kind of distance in inference. The core thesis proposed in this section is that translational trials should never exceed a modest translational distance; to do otherwise threatens the primary objective of the study – scientific utility – as well as the subject's welfare.

The concept of translational distance is intended to structure the ethical analysis of risk and social benefit, much like equipoise. It is intended to prompt researchers, review committees, and policy-makers to contemplate the size of the "inferential gap"[§] separating completed preclinical studies and projected human trial results. Much as a hiker using stones to cross an icy mountain stream will look for the narrowest point because he has the greatest confidence of getting to the other side dry, researchers should seek to minimize the number and extravagance of assumptions in launching human studies. When it proves impossible to cross the stream in a single bound, hikers often achieve dry passage by stone hopping. Clinical investigators confronting a high degree of uncertainty between animal and human studies should analogously contemplate smaller-scale, bridging studies.

For hemophilia gene-transfer trials, reliable large animal models make the task of staying within a modest translational distance straightforward. Because such medically stable volunteers face the risk of opportunity costs and latent adverse events, investigators and IRBs might make further demands on supporting evidence. Cystic fibrosis trials, on the other hand, pose a greater challenge for maintaining a modest translational distance. Rather than conducting a classic translational trial, a preferable option would be to perform a first-in-human trial that tests physiological parameters or validates hypotheses about the investigational intervention. When aggregated with various other modest, "proof-of-principle" human studies, the evidence produced from such studies could potentially justify a phase 1 cystic fibrosis study.

This argument has at least two corollaries. First, it rejects the view, occasionally expressed in the scholarly literature,[26] that first-in-human trials should only be conducted once a "cure" has been demonstrated in animals. Animal evidence will not always be forthcoming, and where it is, it may not be particularly reliable. As noted, a series of modest human studies will often provide sufficient evidence for launching a phase 1 trial. Second, the argument looks askance at those who would argue that delaying translational trials condemns "an undetermined number of patients [to] die who might have been saved if the crucial advances had come sooner."[27] The principle of modest translational distance would place greater emphasis on parallel, modest experiments rather than bold studies that aim to show safety and substantial biological response in a single shot. In so doing, the aims are to enhance the reciprocal

---

[§] I am grateful to Jason Scott Robert for suggesting this phrase.

and iterative value of otherwise "negative" experimental findings, and to maximize the value of any human study.

How, then, might investigators and ethics committees determine whether the threshold of "modest translational distance" has been met? I would suggest that translational distance consists of four components: the internal validity of studies providing supporting evidence, their external validity, the correspondence between preclinical studies and human trial endpoints, and the degree to which independent experts can make reasonably well-informed judgments about each of these components. A study might be said to satisfy conditions of modest translational distance provided all four conditions are fulfilled.

Internal validity and translational distance

Internal validity refers to the ability to make causal inferences about an experimental result. Clinical research has evolved a number of practices – allocation concealment and randomization being but two examples – intended to strengthen internal validity. Adherence to such practices in preclinical research, however, is at best sporadic. Perhaps because neurological interventions have proven especially resistant to translation,[20,21] concerns about methodological rigor are particularly well substantiated in the field of stroke.

For example, a-priori power calculations are virtually unheard of in preclinical studies aimed at showing a biological response. One systematic review of preclinical stroke studies found only one instance in 45 studies in which authors reported performing a-priori power calculations.[28] This has two potential consequences. On the one hand, treatment effects might be missed if studies are inadequately powered. On the other hand, failure to state an a-priori hypothesis, and to power a study accordingly, creates an avenue through which investigators can expand sample sizes or manipulate their definition of efficacy in order to cross a "finish line" of statistical significance.

Another infrequent practice in preclinical research is the use of randomized treatment allocation. Here, the concern is that observed effects might reflect differences between animals receiving the study intervention and those in the control group. Some might question the need for randomizing animals, which, unlike human volunteers, tend to be more genetically homogenous and housed in identical environments. Nevertheless, there are numerous sources of variability among animals used in a study, including sex and body weight. Certain animal populations, like dogs and nonhuman primates, are often genetically outbred. A batch of animals might have an undetected infection that affects outcome.[29] And manipulations such as induction of injuries to simulate a disease process are subject to variation.[¶] Researchers could minimize the effects of such variation through a simple randomization procedure. Yet in one meta-analysis, fewer than half of preclinical studies reported randomization.[28] In other more comprehensive studies that included non-stroke research as well, approximately a third reported randomization.[31]

Blinding provides another way to minimize bias. First, masking investigators to the treatment allocation of their animals reduces the possibility that investigators, subconsciously or not, will stock their control group with sicker animals. Second, blinded (or automated)

[¶] Epidemiologist Ian Roberts explains the need for randomization as follows: "Imagine a cage of 20 rats, and you've got a treatment for some. So you stick your hand in a cage, and pull out a rat. The rats that are the most vigorous are hardest to catch, so when you pull out 10 rats, they're the sluggish ones, the tired ones, they're not the same as the ones still in the cage, and they're the control. Immediately there's a difference between the two groups." See Gawrylewski A. The trouble with animal models.[30]

outcome assessment protects the integrity of measurements involving judgment and skill. Death, for example, is a frequent endpoint in preclinical studies. This might seem a "hard endpoint" whose measure is impervious to biased assessment until one considers that many animal-welfare policies encourage researchers to euthanize moribund animals.[32–34] Such polices, which are amply justified on ethical grounds, introduce a potential for bias in mortality curves that can be reduced significantly if researchers assessing outcome are prevented from knowing whether their animals received active or control interventions. One meta-analysis of 228 preclinical studies showed that only about a fifth assessed outcome blindly.[31]

Do such methodological deficiencies diminish the predictive potential of preclinical studies? Data on this question are conflicting.[35–37] In the largest study to examine this question, researchers pooled the results of thirteen meta-analyses comprising over 15 000 animals in studies of stroke interventions, and examined whether failure to randomize animals and mask investigators led to exaggerated estimates of treatment effects. Perhaps surprisingly, randomization did not have a significant impact on treatment effects. Neither did blinded outcome assessment (though one reason for this could be the use of automated outcome assessment, which would overcome any effect on non-blinding). However, unblinded treatment assignment did produce significantly exaggerated treatment effects.[38]

Studies of methodological rigor in translational oncology, or for specialized platforms like gene transfer, have yet to be performed. Nevertheless, a scan through top-tier journals in either field shows that the methodological deficiencies described above are hardly unique to stroke. Though the jury remains out as to whether these deficiencies bias preclinical studies, the limited data presented above, coupled with logic and evidence of the consequences of bias in human clinical research,[39] argue in favor of strict standards for internal validity in preclinical experiments.

External validity and translational distance

The problem of ensuring a close relationship between preclinical experiments and the clinical setting presents a far more difficult challenge. Though confounders of external validity are virtually limitless, the most perennial one is the animal model.**

Consider oncology studies, which account for the largest volume of gene-transfer trials. Since the mid 1980s, the most widely used system in preclinical testing has been the murine xenograft of human tumor tissue. In this model, scientists implant human tumor tissues into the mouse, typically underneath the skin (ectopically), but sometimes to anatomical locations that investigators are attempting to model (orthotopically). To prevent rejection of grafts, researchers use mice that lack functional immune systems. Once the tumor reaches a certain size, investigators then administer the experimental agent and test for tumor response.

The discontinuities between mouse xenograft models and human cancer patients are myriad. First, mouse tumors are one or two orders of magnitude smaller than human tumors; as such, they are much less likely to develop characteristics, like hypoxia or clonal variation, that are typical of human tumors. Second, tumor grafts are often less invasive than spontaneous human cancers, and have a different morphology. Third, immunodeficient mice will not reflect how a human volunteer's immune system might modulate the effects of a study agent (a critical factor, given that many gene-transfer strategies involve immunogenic vectors). Fourth, xenograft

** Many important translational studies, including Alain Fischer's X-SCID experiment, proceeded on the basis of human rather than animal evidence; however, for the sake of brevity, I will concentrate my discussion on animal models.

hosts are inbred, and therefore do not reflect the genetic diversity of subjects in clinical testing. Last, tumor tissues are typically propagated in vitro for many generations before implantation; as a result, they acquire properties that differ from those of native tumor tissues.[40]

Several retrospective analyses make some sense of the predictive value of tumor xenograft models. On a positive note, all cancer drugs found to be effective against human cancers have shown activity in mouse xenografts (thus the worry that failing to cure mice might lead researchers to abandon promising drug candidates appears unfounded).[41] On the other hand, xenograft models produce false positives at a disturbingly high rate. One study found that fewer than half of all agents showing activity in xenograft models were clinically active in human phase 2 studies[42] (most cancer researchers I speak with believe this figure is optimistic).

Diseases for which animal models do not present formidable external validity problems are largely confined to a small set of single-gene disorders that occur spontaneously in large animals. So what is to be done about this problem? Several approaches might, in the long run, offer effective ways of bridging the translational divide between animals and human beings (see Box 7.2).

Box 7.2 Bridging translational distance: different models and trial designs

In their 1995 report to the NIH on gene transfer, the Orkin–Motulsky panel recommended "increased emphasis on ... further development of animal models of disease." Returning to the example of cancer, several improved models are being explored. Various groups, for example, have attempted to genetically engineer mice to develop various types of "spontaneous" tumor. Though still in the early stages of development, transgenic models overcome many limitations of the xenograft model like the lack of a functional immune system.[40] Another emergent approach is the enrollment of companion animals in novel intervention studies. Dogs, it turns out, develop many cancers at approximately the same rate as human beings.[43] At a cancer incidence of two in a hundred[44] and a total population of 65 million in the USA, pet dogs provide a pool of 1.3 million candidates for translational cancer studies. Several novel interventions have been tested in companion animals before advancing into human beings[43] (one illustrious example is angiogenesis inhibitors[45]), and in 2003, the US National Cancer Institute established a comparative oncology program to facilitate such studies. Companion animals offer some obvious advantages in terms of external validity: they are immunocompetent, genetically diverse, and their tumors are large and spontaneous. Companion animals might have several ethical advantages over experimentally induced cancer models – though they also present ethical challenges with respect to review and consent.[46]

A second way of complying with the principle of modest translational distance where animal models are wanting would be the use of pilot and feasibility studies in human beings. These have the advantage of gathering critically needed information about the properties of an intervention without exposing volunteers systemically to a study agent. Thus, several cystic fibrosis gene-transfer interventions were initially studied in human beings by testing physiological parameters, such as potential difference across modified tissues, after delivery of the agent to the nasal epithelia.[47] This enabled investigators to determine whether human tissues could express the transgene, and whether the transgene was functional, without administering large doses of agent to vulnerable and relatively inaccessible lung tissue. In cancer as well as other areas of clinical research, exploratory trials (elsewhere referred to as "phase 0" and "microdosing" studies) provide another way of testing biological hypotheses before exposing volunteers to large quantities of experimental agent. These studies involve administering very small doses of an experimental agent in order to estimate pharmacokinetics, biodistribution, or biological activity. Such studies are ethically akin to the cystic fibrosis studies described above in that they minimize risk while enabling researchers to gather intelligence that might inform further refinement in the laboratory, or a smarter phase 1 study design.[48]

Correspondence between preclinical and clinical studies

A third factor closely related to external validity is the correspondence between the goals of the human study and the experiments supporting it. That is, preclinical studies might have high internal and external validity, but translational distance might nevertheless be great because of mismatches between the human protocol and its supporting experiments. First, the human study might use strategies that depart from those used in preclinical studies. For instance, if investigators pursue preclinical studies by delivering vector to a particular anatomical location, they should refrain from performing the human study by delivering it to a different anatomical location.

Second, correspondence is diminished wherever human studies aim to test hypotheses that are either not validated in preclinical models or not strongly supported by interpolation of results from various preclinical and clinical studies. For example, blood serum factor IX levels above 1 percent are sufficient to convert severe hemophilia B into a moderate form. The correspondence between supporting evidence and a human protocol would be violated if supporting experiments only achieve 0.75 percent, but the investigators are aiming for therapeutic levels. Similarly, if investigators set out to establish that their vector can correct coagulant factor deficiency out to one year, their supporting evidence should show correction to one year as well.

Once again, these recommendations might seem brainless. Yet mismatch between preclinical and clinical studies is a not infrequent occurrence with first-in-human studies. Some notable examples include Cline's ß-thalassemia protocol (animal studies did not show correction),[49] the adenoviral OTC study (recall from Chapter 3 that investigators switched to third-generation adenoviral vectors after having performed their preclinical studies in second-generation vectors), and a Parkinson's disease study injecting AAV vectors encoding the neurturin gene (according to RAC minutes, investigators proposed to inject their vectors into the striatum when their preclinical efficacy data involved injection into the substantia nigra).[50]

In some instances, investigators might switch techniques or plans on the basis of information acquired between the completion of preclinical testing and the initiation of human studies. Thus, for example, Wilson's substitution of newer adenoviral vectors was premised on what seemed a reasonable judgment that newer vectors conferred a higher degree of safety. There are almost certainly instances where such protocol modifications are justified. Yet they create challenges for interpreting results. More broadly, they signal that the knowledge environment surrounding a particular protocol is highly unstable. Researchers should proceed especially cautiously in such circumstances, lest a trial become scientifically obsolete before completion.

The appraisal of credibility

Translational distance relates to the degree of confidence among relevant experts that laboratory observations predict clinical outcomes. As noted in the previous three sections, this confidence will depend crucially on various characteristics of the supporting evidence and protocol. But how confident can we be that reviewers have all the information they need to assess translational distance? And more broadly, to what degree can they trust the information with which they are provided?[††]

[††] In a provocative article, epidemiologist John Ioannidis provides six factors that diminish the probability that a research finding is false: (1) small sample size, (2) small effect size, (3) the greater the number

One major concern in translational research is optimism bias. This refers to a conscious or subconscious tendency of investigators to present their data in a favorable light. According to some commentators, something akin to optimism bias might pose particular challenges for targeted and rational therapeutic translational research. At the point where preclinical studies are under way, investigators have often devoted many years to establishing a biological hypothesis. Though, as noted in Chapter 3, personal identification with a strategy might fuel perseverance, it can also interfere with a dispassionate appraisal and presentation of study findings.[52]

Two channels through which optimism bias might find expression – but that might not be apparent in the review of supporting evidence – are the management of data outliers and missing data. The former refers to measurements that significantly depart from all others in a study. Because sample sizes in animal studies are usually modest, the decision to include or exclude these measurements can have major consequences for the statistical analysis of preclinical experimental results. The problem of missing data arises when researchers exclude measurements because of a technical problem (e.g. measurements are not made because of equipment failure), a confounding issue (e.g. an animal's response is not assessed because it developed a concurrent illness), or some other SNAFU (e.g. the technician couldn't get to the lab because of a massive snowstorm). How such "missing data" are addressed and interpreted can also have significant effects on a study's outcome.[53]

Another important set of considerations in assessing credibility is financial interest. Translational clinical investigators are frequently named as co-inventors on patents relating to the study intervention. For example, one recent study found that 35 percent of protocols submitted to the RAC between the years 1999 and 2002 involved principal investigators who were named as inventors.[54] Under the US Bayh-Dole act, inventors are entitled to a portion of royalties earned on such patents.[55] Although the effects of patent interests on the conduct of investigators have not been studied empirically, the suggestion that they might have consequences with respect to bias and information non-disclosure is at least a plausible one.[56] Another financial issue concerns milestone payments. Biotechnology companies and universities often receive payments from investors at key junctures during the development of new products. One is the filing of an investigational new drug application with the FDA. Between 1980 and 2003, universities earned \$17 million from milestone payments occurring during preclinical and phase 1 testing of biotechnology products.[57] Such payments can encourage premature entry into clinical testing.

Publication bias presents another challenge to the credibility of preclinical experiments. Numerous studies have shown that, until the advent of clinical trial registries, as many as 40 percent of randomized controlled trials went unpublished, with lack of treatment effect being a significant predictor of non-publication. The extent of publication bias in preclinical research is less well documented. Nevertheless, several studies have statistically shown a high likelihood of publication bias (by demonstrating strong relationships between small samples sizes and large treatment effects[31,35,58]), and the withholding of negative results is likely to be a frequent occurrence in preclinical efficacy studies.

Several fairly straightforward policies might enhance the credibility of supporting evidence claims. First, preclinical studies should include a detailed discussion of limitations and alternative explanations for observed phenomena (T. Ramsay, personal communication).

of tested relationships, (4) the greater the flexibility in designs, definitions, and outcomes, (5) greater financial and professional interests, (6) more fashionable research areas.[51]

Reviewers will often lack either the time or expertise to perform a meticulous analysis of a preclinical dataset. A description of study limitations might help direct the attention of reviewers to possible weaknesses in supporting evidence. Second, first-in-human proposals should be accompanied by systematic reviews of related interventions in human beings. This would enable referees to evaluate the supporting evidence in the context of other related studies.[59,60] Third, preclinical studies should include a detailed disclosure of missing data, and a discussion of how missing data and outliers were managed during analysis. Fourth, investigators should disclose all financial interests relating to a study intervention, including patents, consultancies, and milestone payments linked to IND filing and/or trial initiation.

Another set of recommendations for establishing the credibility of supporting evidence is likely to be more controversial. Some commentators have proposed that preclinical studies be prospectively registered on public databases.[31,61] Such a policy would encounter strong opposition from investigators and biotechnology firms, since it could divulge proprietary and strategic information. Investigators would likely further object that registries impose rigidity on the research process. Nevertheless, there might be compelling reasons to favor registries in research areas where publication bias is likely to be extensive, or where preclinical experiments might be ethically contentious (for example, because they necessarily inflict serious suffering on non-human primates). Alternatively, researchers might establish data repositories to pool preclinical studies.[51]

Investigators might also be expected to publish their preclinical data in full prior to the initiation of a human study. This would provide an opportunity for the relevant peer community to evaluate the quality and quantity of evidence; it would also offer a modicum of assurance that the strategy has been vetted. IRBs might further ask investigators to submit manuscript referee comments along with their protocol. More controversial still would be a requirement that key preclinical studies be replicated by independent and financially disinterested research teams prior to the initiation of human studies. This would serve as a check on optimism bias. Of course, such a proposal could significantly increase expense and slow entry into clinical testing. On the other hand, independent verification of core claims might be sought in special circumstances, for example where first-in-human studies necessarily involve an unusual degree of risk.

Applying this analysis to the AAV-GAD protocol

In summary, then, maintaining modest translational distance provides greater confidence that the risks of a first-in-human study are appropriately balanced by their potential scientific value. An appraisal of translational distance requires assessment of the internal and external validity of preclinical studies, the correspondence between what was established in supporting experiments and what the investigators intend to show in their human study, and assurance that study teams have taken reasonable measures to manage optimism bias and to secure the credibility of supporting evidence (see Table 7.1).

To what degree might Kaplitt's Parkinson's disease protocol fulfill these requirements? One should be very cautious about venturing retrospective judgments, especially without having access to the full RAC submission. Nevertheless, published preclinical experiments and RAC deliberations allow a tentative appraisal.

Was the study risky enough to warrant a high evidentiary threshold for initiation? The risks of the study divided into two major components: the administration of the agent (which requires insertion of a needle to deep brain structures) and the agent itself. The former is a "baseline" risk: the probability and magnitude of harms are more or less independent of the

Table 7.1 Translational distance: sources and control measures

Factor	Concerns	Measures for reducing translational distance
Internal validity	Poor methodological rigor	a priori power calculation
		a priori statement of hypothesis
		Randomization
		Blinded treatment allocation
		Blinded outcome assessment
External validity	Poor animal model	Use of alternative animal models (e.g. companion animals)
		Test of hypothesis in small-scale, limited human study (e.g. phase 0 trials)
Correspondence	Non-validated hypothesis	Hypothesis tested in human studies corresponds to that used in animal studies
	Non-validated methods	Close adherence in human studies to experimental conditions used in animal studies
Credibility	Optimism bias	Full discussion of limitations of supporting data; alternative explanations for observed effects; discussion of missing data
	Non-publication	Publication of negative results; publication of effect studies before initiating human trials; prospective registration of preclinical effect studies

dose or properties of the experimental agent. Based on experience of implanting electrodes for deep-brain stimulation, the risk of intracerebral hemorrhage leading to serious and/or permanent disability is 0.5–1 percent.[62,63] Risks of the study agent are impossible to estimate with any great confidence. The investigators provided some assurance that they had never once observed untoward consequences in preclinical studies. However, the team's only published study of administering the agent to non-human primates that had Parkinsonian symptoms involved only 13 macaques. At that sample size, the study team had 95 percent power of detecting a serious adverse event that occurs about once in every five animals.[‡‡] That is, only *extremely* frequent serious adverse events were statistically excluded in preclinical testing. These risk levels are extraordinary when one considers that they are not ethically justified by an appeal to their therapeutic value (see Chapter 5). Friedmann's admonishment that investigators support their study "to the hilt" therefore seems well justified.

‡‡ A quick formula for calculating the ability to detect adverse events is $x = -1.3/n$, where n = sample size. If you then calculate the antilog of x, you will have the proportion of adverse events that would have to occur in order to have a 95 percent power for detecting it. Thus, to have 95 percent power for detecting an adverse event that occurs once in a hundred times, the researcher would need a sample size of about 300 animals (or human beings). To have 95 percent power to detect an adverse event occurring with a frequency of one in ten, the sample size needed is closer to 30. See Dell *et al.*[64]

Did the study cross a modest translational distance? The human protocol was supported by three preclinical studies – two in rats, and one in non-human primates. None reported blinded treatment allocation or a-priori power calculations. Two reported the use of blinded outcome assessment. One randomized animals. The preclinical studies do not appear to have met unusually high standards of internal validity.

For reasons previously described, external validity presents a major problem for Parkinson's disease research. Perhaps the field could be more enterprising about developing better models. For example, one commentator has proposed several alternatives that might better recapitulate the chronic and degenerative character of Parkinson's disease in human beings.[65] I will nevertheless continue my analysis on the premise that the value of such models has not yet been established. I argued previously that, when human disease and animal models are very different, investigators should consider a series of smaller, pilot-scale studies before embarking on a full phase 1 study that exposes volunteers to biologically active doses of a novel agent. Because of the significant "fixed" risk of administration and the anatomical specificity of the intervention, however, it is very difficult to think of how any pilot or microdose study could significantly improve the risk–value balance. As such, Kaplitt's protocol could very well count as one of those exceptional circumstances where attainment of "modest translational distance" by performing pilot studies might have been ill-advised. Nevertheless, at the time investigators proposed their study to the RAC, they had collected only preliminary supporting evidence from non-human primates. The team did, however, publish their macaque studies a few months before the completed human study appeared in *Lancet*. These used an experimental model involving acute and relatively mild disease. The model has advantages in terms of reproducibility and expedience.[66] At the same time, the study did not explain why the acute model, which does not show hallmark features of human illness like dyskinesias, was chosen instead of a chronic disease model.[67] Another concern in this study was the inexplicable appearance of necrotic lesions in certain brain regions of the control animals. Might this have resulted in an exaggerated treatment effect?

Perhaps even more troubling are the problems with correspondence. The macaque study did not show statistically significant efficacy between experimental and control arms on a clinical rating scale. Yet the human study was a classic phase 1 trial aimed primarily at establishing a safe and effective dose for phase 2 studies. On what basis was the strategy expected to work in human beings when it hadn't proven effective in non-human primates?

Finally, were preclinical data presented to the relevant expert community in a way that inspires confidence and credibility? On the one hand, their approach has been validated in rats by an independent research team.[68] And the lead investigators, who held significant financial interests relating to the study intervention, seem to have been forthright with disclosure. On the other hand, the independent replication was not published until well after the human trial received RAC approval. Nor, as noted above, were non-human primate data.

In summary, the proposal appears to have exceeded a modest translational distance. This is troubling in light of the significant risk involved in the protocol. The investigators might have substantially reduced translational distance by improving the internal validity of their preclinical studies, using more appropriate animal models, obtaining more significant results in animals, and providing greater transparency by publishing their supporting evidence before trial initiation. Because of the risk associated with accessing targeted tissue, pilot or exploratory studies might not have been a viable option for reducing translational distance. Still, one design modification that might have improved the risk–benefit balance for all but the first volunteer would have been to wait longer than a month or two between dosing volunteers. After all, subjects receiving embryonic dopamine neuron transplants did not develop dyskinesias until a year after surgery.[69]

Conclusions

This chapter argued that first-in-human studies are likely to have greater scientific value when they are buttressed by strong supporting evidence. Riskier protocols should generally be held to more stringent evidentiary standards. Nevertheless, because of their high degrees of indeterminacy about risk, all first-in-human trials of novel interventions should satisfy the condition of modest translational distance by assuring high standards of internal and external validity for preclinical experiments, close correspondence between preclinical studies and clinical trial protocols, and transparency surrounding data management, professional interests, and financial relationships in the study intervention.

The proposal that all novel intervention first-in-human trials be held to a common standard of translational distance might face several objections. Some might find my proposal restrictive, paternalistic, and excessively cautious. A skeptic might ask, why not allow desperate human volunteers greater flexibility to choose risk levels? Though I agree that volunteers should play a more significant role in defining and assessing risk and clinical value, if ever there were a place for paternalism in research ethics, evaluation of preclinical evidence is it: who but the most elite is qualified to absorb and weigh the complex and numerous preclinical research associated with a given protocol?

Some might worry that my proposal would hold back medical innovation, or that it would stifle the restless creativity of medical researchers like During and Kaplitt. This is indeed a cautionary proposal, and in some instances, the standard advanced here might prove inhibitory. But I think that there are several responses to this charge. First, the proposal is aimed at maximizing the ratio of scientific value to risk for first-in-human studies. Second, many commentators lament the disappointing pace of translating basic science discoveries into clinical applications. For instance, noting the very high failure rate for translating basic discoveries into clinical applications, epidemiologist John Ioannidis has urged greater replication, better study design, focus on credibility, robust methodology, and systematic synthesis of data.[70] In 1999, academic researchers and industry representatives began convening a series of meetings directed at improving success rates for translating neurological interventions. Recommendations stemming from their first policy statement, STAIR I, include: randomization and blinding in preclinical studies, matching preclinical models to the human disease state, and independent replication.[71] Last, would anyone seriously argue that we ought to give clinical investigators a pass when it comes to internal validity, or matching their trial designs to preclinical studies?

In another vein, skeptics might argue that the 2007 *Lancet* publication of their human trial vindicates the safety and plausibility of their strategy. But as Henry Beecher once famously said, "a human study is ethical or not at its inception; it does not become ethical *post hoc*."[72]

The concept of "translational distance" would seem to suggest an objective metric, or at least a commonly agreed upon standard. Yet my account leaves vague exactly how translational distance might be defined and measured. One response to this criticism is that many seemingly mathematical concepts in bioethics have a metaphorical character. The concept of risk, which is often defined by the product formula, is one example. Few have credibly argued that the product formula provides a mathematical way of rendering commensurable the diverse burdens and risk experiences volunteers might encounter. Similarly, "appropriate risk–benefit balance" and "equipoise" both lack metric precision.

Nevertheless, the bank of experience for assessing risk and equipoise is clearly more developed than for translational distance. How might appropriate translational distance be better defined? Two approaches might be explored. Expert practitioners might be surveyed about their degree of confidence that study investigators have adequately addressed each of the four components of translational distance. Composite scores above a certain level might be taken to indicate that a proposal exceeds an acceptable translational distance. Alternatively, modest translational distance might be benchmarked by constructing a description of protocols that, on the basis of consensus opinion among expert practitioners, fall within a modest translational distance. Assessments of future protocols might then be made inductively by comparison with standards.

As both approaches would suggest, translational distance is not an objective quantity that can stand apart from a community of experts. The framework described in this chapter thus aims at providing a structure that can be mapped to procedural and formalized approaches for deciding trial initiation. In gene transfer as with infrastructure planning, finding the narrowest, most useful point to span a chasm is a community effort.

References

1. Recombinant DNA Advisory Committee (RAC). Minutes of Meeting June 14–15, 2001.
2. Grady D, Kolata G. Gene therapy used to treat patients with Parkinson's. *New York Times* 2003 August 19.
3. Edelman S. Brain-gene op may flop; Parkinson's docs rip 'risky' surgery. *New York Post* 2003 September 21.
4. King NM, Cohen-Haguenauer O. En route to ethical recommendations for gene transfer clinical trials. *Mol Ther* 2008; **16**(3): 432–8.
5. Kaplitt MG, Friedmann T. Gene therapy for Parkinson's disease. National Public Radio. *Talk of the Nation*, August 22, 2003.
6. US Food, and Drug Administration. *Guidance for Industry. INDS – Approaches to Complying with CGMP During Phase 1* (Draft Guidance). US Department of Health and Human Services, 2006.
7. Freedman B. Equipoise and the ethics of clinical research. *N Engl J Med* 1987; **317**(3): 141–5.
8. Anderson J, Kimmelman J. Equipoise in First-in-Human Clinical Trials: A Bridge Too Far? submitted.
9. Djulbegovic B. Articulating and responding to uncertainties in clinical research. *J Med Philos* 2007; **32**(2): 79–98; at 84.
10. London AJ. Clinical Equipoise: Foundational Requirement or Fundamental Error? In: Steinbock B, ed. *The Oxford Handbook of Bioethics*. New York, Oxford University Press, 2007; 571–96.
11. Anderson WF, Fletcher JC. Sounding boards. Gene therapy in human beings: when is it ethical to begin? *N Engl J Med* 1980; **303**(22): 1293–7.
12. Rood PP, Cooper DK. Islet xenotransplantation: are we really ready for clinical trials? *Am J Transplant* 2006; **6**(6): 1269–74.
13. Svendsen CN, Langston JW. Stem cells for Parkinson disease and ALS: replacement or protection? *Nat Med* 2004; **10**(3): 224–5.
14. King NM. RAC oversight of gene transfer research: a model worth extending? *J Law Med Ethics* 2002; **30**(3): 381–9.
15. Halpern SA. *Lesser Harms: The Morality of Risk in Medical Research*. Chicago, University of Chicago Press, 2004.
16. Fox RC, Swazey JP. *The Courage to Fail: A Social View of Organ Transplants and Dialysis*. New Brunswick, Transaction Publishers, 2001.
17. McCarthy C. Justice in federal research policy. In: Kahn JP, Mastroianni AC, Sugarman J, eds. *Beyond Consent: Seeking Justice in Research*. New York, Oxford University Press, 1998; 17
18. Nuremberg Code. In: *Trials of War Criminals before the Nuremberg Military Tribunals under Control Council Law No. 10, Vol. 2*. Washington, DC, US Government Printing Office, 1949; 181–2.
19. Council for International Organizations of Medical Sciences (CIOMS). International ethical guidelines for biomedical research

involving human subjects. *Bull Med Ethics* 2002; **182**: 17–23. Commentary for Guideline 8.

20. Kola I, Landis J. Can the pharmaceutical industry reduce attrition rates? *Nat Rev Drug Discov* 2004; **3**(8): 711–15.
21. Pangalos MN, Schechter LE, Hurko O. Drug development for CNS disorders: strategies for balancing risk and reducing attrition. *Nat Rev Drug Discov* 2007; **6**(7): 521–32.
22. Hackam DG, Redelmeier DA. Translation of research evidence from animals to humans. *JAMA* 2006; **296**(14): 1731–2.
23. Orkin SH, Motulsky AG. Report and Recommendations of the Panel to Assess the NIH Investment in Research on Gene Therapy. US National Institutes of Health, Office of Biotechnology Activities, 1995; 10.
24. Johnson JR, Williams G, Pazdur R. End points and United States Food and Drug Administration approval of oncology drugs. *J Clin Oncol* 2003; **21**(7): 1404–11.
25. Pinch T. One, two, three … testing!: Toward a sociology of testing. *Science, Technology & Human Values* 1993; **18**(1): 25–41.
26. Fiester A, Scholer H, Caplan A. Stem cell therapies: time to talk to the animals. *Cloning Stem Cells* 2004; **6**(1): 3–4.
27. Mercola KE, Cline MJ. Sounding boards. The potentials of inserting new genetic information. *N Engl J Med* 1980; **303**(22): 1297–300.
28. van der Worp HB, de Haan P, Morrema E, Kalkman CJ. Methodological quality of animal studies on neuroprotection in focal cerebral ischaemia. *J Neurol* 2005; **252**(9): 1108–14.
29. Dirnagl U. Bench to bedside: the quest for quality in experimental stroke research. *J Cereb Blood Flow Metab* 2006; **26**(12): 1465–78.
30. Gawrylewski A. The trouble with animal models: Why did human trials fail. *The Scientist* 2006; **21**(7): 44.
31. Perel P, Roberts I, Sena E, Wheble P, Briscoe C, Sandercock P, *et al.* Comparison of treatment effects between animal experiments and clinical trials: systematic review. *BMJ* 2007; **334**(7586): 197–202.
32. Australian Government. *Australian Code of Practice for the Care and Use of Animals for Scientific Purposes*, 7th edn. National Health and Medical Research Council, 2004; at 1.27.
33. Organization for Economic Cooperation and Development. No. 19. *Guidance Document on the Recognition, Assessment, and Use of Clinical Signs as Humane Endpoints for Experimental Animals Used in Safety Evaluation.* Paris, Environment Directorate, 2000; 20.
34. UK Co-ordinating Committee on Cancer Research. *UKCCCR Guidelines for the Welfare of Animals in Experimental Neoplasia*, 2nd edn. London, UKCCCR, 1997; at 3.4.
35. Macleod MR, O'Collins T, Horky LL, Howells DW, Donnan GA. Systematic review and metaanalysis of the efficacy of FK506 in experimental stroke. *J Cereb Blood Flow Metab* 2005; **25**(6): 713–21.
36. Macleod MR, O'Collins T, Horky LL, Howells DW, Donnan GA. Systematic review and meta-analysis of the efficacy of melatonin in experimental stroke. *J Pineal Res* 2005; **38**(1): 35–41.
37. Bebarta V, Luyten D, Heard K. Emergency medicine animal research: does use of randomization and blinding affect the results? *Acad Emerg Med* 2003; **10**(6): 684–7.
38. Crossley NA, Sena E, Goehler J, Horn J, van der Worp B, Bath PM, *et al.* Empirical evidence of bias in the design of experimental stroke studies: a metaepidemiologic approach. *Stroke* 2008; **39**(3): 929–34.
39. Kunz R, Vist G, Oxman AD. Randomisation to protect against selection bias in healthcare trials. *Cochrane Database Syst Rev* 2007(2): MR000012.
40. Becher OJ, Holland EC. Genetically engineered models have advantages over xenografts for preclinical studies. *Cancer Res* 2006; **66**(7): 3355–8; discussion 3358–9.
41. Sausville EA, Burger AM. Contributions of human tumor xenografts to anticancer drug development. *Cancer Res* 2006; **66**(7): 3351–4; discussion 3354.
42. Kelland LR. Of mice and men: values and liabilities of the athymic nude mouse model in anticancer drug development. *Eur J Cancer* 2004; **40**(6): 827–36.
43. Hansen K, Khanna C. Spontaneous and genetically engineered animal models; use in preclinical cancer drug development. *Eur J Cancer* 2004; **40**(6): 858–80.

44. Dobson JM, Samuel S, Milstein H, Rogers K, Wood JL. Canine neoplasia in the UK: estimates of incidence rates from a population of insured dogs. *J Small Anim Pract* 2002; **43**(6): 240–6.
45. Rusk A, McKeegan E, Haviv F, Majest S, Henkin J, Khanna C. Preclinical evaluation of antiangiogenic thrombospondin-1 peptide mimetics, ABT-526 and ABT-510, in companion dogs with naturally occurring cancers. *Clin Cancer Res* 2006; **12**(24): 7444–55.
46. Kimmelman J, Nalbantoglu J. Faithful companions: a proposal for neurooncology trials in pet dogs. *Cancer Res* 2007; **67**(10): 4541–4.
47. Knowles MR, Paradiso AM, Boucher RC. In vivo nasal potential difference: techniques and protocols for assessing efficacy of gene transfer in cystic fibrosis. *Hum Gene Ther* 1995; **6**(4): 445–55.
48. Kimmelman J. Ethics at phase 0: clarifying the issues. *J Law Med Ethics* 2007; **35**(4): 727–33.
49. Beutler E. The Cline affair. *Mol Ther* 2001; **4**(5): 396–7.
50. Recombinant DNA Advisory Committee (RAC). Minutes of Meeting March 16, 2005; 9.
51. Ioannidis JP. Why most published research findings are false. *PLoS Med* 2005; **2**(8): e124.
52. Lindner MD. Clinical attrition due to biased preclinical assessments of potential efficacy. *Pharmacol Ther* 2007; **115**(1): 148–75.
53. Unger EF. All is not well in the world of translational research. *J Am Coll Cardiol* 2007; **50**(8): 738–40.
54. Durell KL, Kimmelman J, Nalbantoglu J, Gold ER. *Patent Inventorship of Principal Investigators in Phase I Gene Transfer Clinical Trials: An Empirical Analysis*. Conference Proceedings, 10th Annual Meeting of the American Society of Gene Therapy, Seattle, WA, June 2 2007.
55. 35 U.S.C. § 202.(c)(7)(B).
56. Kimmelman J. Inventors as investigators: the ethics of patents in clinical trials. *Acad Med* 2007; **82**(1): 24–31.
57. Edwards MG, Murray F, Yu R. Value creation and sharing among universities, biotechnology and pharma. *Nat Biotechnol* 2003; **21**(6): 618–24.
58. Benatar M. Lost in translation: treatment trials in the SOD1 mouse and in human ALS. *Neurobiol Dis* 2007; **26**(1): 1–13.
59. Moher D, Schulz KF, Altman DG. The CONSORT statement: revised recommendations for improving the quality of reports of parallel-group randomised trials. *Lancet* 2001; **357**(9263): 1191–4.
60. Chalmers I, Matthews R. What are the implications of optimism bias in clinical research? *Lancet* 2006; **367**(9509): 449–50.
61. Hackam DG. Translating animal research into clinical benefit. *BMJ* 2007; **334**(7586): 163–4.
62. Sansur CA, Frysinger RC, Pouratian N, Fu KM, Bittl M, Oskouian RJ, *et al.* Incidence of symptomatic hemorrhage after stereotactic electrode placement. *J Neurosurg* 2007; **107**(5): 998–1003.
63. Binder DK, Rau GM, Starr PA. Risk factors for hemorrhage during microelectrode-guided deep brain stimulator implantation for movement disorders. *Neurosurgery* 2005; **56**(4): 722–32.
64. Dell RB, Holleran S, Ramakrishnan R. Sample size determination. *ILAR J* 2002; **43**(4): 207–13.
65. Bezard E. A call for clinically driven experimental design in assessing neuroprotection in experimental Parkinsonism. *Behav Pharmacol* 2006; **17**(5–6): 379–82.
66. Emborg ME. Nonhuman primate models of Parkinson's disease. *ILAR J* 2007; **48**(4): 339–55.
67. Emborg ME, Carbon M, Holden JE, During MJ, Ma Y, Tang C, *et al.* Subthalamic glutamic acid decarboxylase gene therapy: changes in motor function and cortical metabolism. *J Cereb Blood Flow Metab* 2007; **27**(3): 501–9.
68. Lee B, Lee H, Nam YR, Oh JH, Cho YH, Chang JW. Enhanced expression of glutamate decarboxylase 65 improves symptoms of rat parkinsonian models. *Gene Ther* 2005; **12**(15): 1215–22.
69. Freed CR, Greene PE, Breeze RE, Tsai WY, DuMouchel W, Kao R, *et al.* Transplantation of embryonic dopamine neurons for severe Parkinson's disease. *N Engl J Med* 2001; **344**(10): 710–19.

70. Ioannidis JP. Evolution and translation of research findings: from bench to where? *PLoS Clin Trials* 2006; **1**(7): e36.
71. Recommendations for standards regarding preclinical neuroprotective and restorative drug development. *Stroke* 1999; **30**(12): 2752–8.
72. Beecher HK. Ethics and clinical research. *N Engl J Med* 1966; **274**(24): 1354–60.

Chapter 8

Tropic of cancers: gene transfer in resource-poor settings

Introduction

Adenosine deaminase-deficient severe combined immune deficiency (ADA-SCID) is a parent's nightmare. The disease is caused by a deficiency in an enzyme responsible for breaking down toxic metabolites that would otherwise kill lymphocytes. Untreated, infants develop recurrent infections and fail to put on weight; they rarely live past two.

But to scientists trying to establish proof of principle for gene transfer, ADA-SCID is a dream: only small amounts of gene correction are needed to restore immune function, tissues are easy to procure for genetic modification, and theoretically, treated cells should have a survival advantage over those that are not corrected. The attractiveness of ADA-SCID is so high, and its incidence so low (approximately one case per 100 000 births)[2] that researcher Stuart Orkin once commented "more [gene transfer researchers would soon be] working on ADA deficiency than there are patients who have it."[3]

But there's a catch. Since the late 1980s, a relatively safe and effective enzyme replacement therapy – PEG-ADA – has been available for ADA-SCID patients. Denying PEG-ADA to children in gene-transfer trials would be unethical, because it would expose them to the risk of relapse. Yet concurrent treatment with PEG-ADA would have scientific costs: it would confound interpretation of the subjects' responses (if volunteers improved, how would investigators be able to tell whether this was owing to the gene transfer or the enzyme replacement?) and it would sustain uncorrected cells (thus dampening the potential therapeutic effects of gene transfer).

Throughout the 1990s, several teams attempted ADA-SCID gene transfer. Each, however, used concurrent enzyme replacement therapy and none produced clear therapeutic successes. Towards the end of the decade, however, a group of researchers in Milan hit on an approach that seemed to overcome ethical concerns about denying standard care. PEG-ADA treatments cost between $200 000 and $300 000 each year.[4] What if the researchers recruited patients from countries that were unable to afford PEG-ADA?

In 2002, the Milan team reported successful restoration of immune function in two infants, one from Colombia, the other from Palestine.[5] Their results are celebrated as one of the first, unambiguous clinical successes in gene transfer. The study would seem a humanitarian "win-win": it made a significant scientific and medical advance against hereditary blood diseases, while saving impoverished infants facing nearly certain death. But to others, the Milan study might have seemed an ethical "perfect storm:" researchers in a highly visible, still somewhat controversial field recruit desperately ill, economically deprived children to participate in a study that would otherwise have been unethical to perform in patients from Europe or the USA. So which is it: are studies like the Milan ADA-SCID trial cases of scientists off loading

This chapter draws heavily on a paper I published with Alex John London.[1]

risky research to impoverished communities, or are they better characterized as researchers doing well by doing serious good?[6]

Factors that motivate early-phase trials in resource-poor settings

The ADA-SCID experiment might appear to be an isolated example of translational trials seeking subjects from low- or middle-income countries (LMICs).[†] Yet, as early as 1992, Kenneth Culver – one of Anderson's collaborators on the first ADA-SCID study – cited the needs of "underdeveloped" countries to justify research on gene-transfer strategies against cancer.[7] Though recruitment of subjects from impoverished settings is by no means standard, many other instances can be cited. These classify into five categories.

Fortuity: economically underprivileged persons sometimes provide investigators with opportunities that would not be available in the sponsoring country. The Milan ADA-SCID study is one clear example: such a trial would not have been ethically permissible in a high-income country (HIC). One important characteristic of persons from LMICs that provides fortuitous research opportunities to HIC researchers is their "treatment naivety" – that is, their lack of access to other treatments that could confound interpretation of the results or otherwise interfere with a meaningful trial.

Expedience: recruitment is often rate limiting in clinical trials. This is particularly true where disorders are rare, and even more so where disorders are rare but medically manageable. Expanding recruitment to other, more populous countries helps ensure timely completion of a study. One gene-transfer trial that appears to have sought volunteers from LMICs because of expedience is Avigen's study of AAV vectors for treating hemophilia B.[8] The study primarily involved North American subjects, but several volunteers were recruited from Brazil; according to clinical trial databases, the trial also recruited in India,[9] and an ongoing trial (as of this writing) is recruiting volunteers in Brazil.[10]

Ultra-rarity: some diseases occur with such low incidence that no country can provide sufficient numbers of subjects. For example, Genzyme corporation is one of the most successful manufacturers of orphan drugs (that is, drugs used to treat diseases that are not a major focus for the pharmaceutical industry). Some of their products involve a class of rare genetic disorders, lysosomal storage diseases. Genzyme has reportedly enrolled subjects from countries like Palestine and Peru.[11] Within gene transfer, Washington University of St Louis researchers are pursuing gene transfer against mucopolysaccharidosis type VII (also known as MPS-VII or Sly syndrome). As of 2002, only 100 cases of the disease had been documented worldwide,[12] and incidence has been estimated at less than one case in two million.[13] Finding enough subjects for the MPS-VII study has necessitated a global approach; at least three subjects have been recruited from Brazil.[14]

Prevalence: some diseases, like HIV-AIDS, hepatitis C, malaria, and β–thalassemia, occur with much greater frequency in LMICs. Thus, sponsors like the National Institute of Allergy and Infectious Diseases, Merck, and VIRxSYS have tested a number of gene-transfer HIV vaccines in early-phase trials in Africa and India. Some of these used used adenovirus[15,16] and AAV vectors,[17] others naked DNA,[18] and still others involved one of the earliest applications of lentivirus-based vectors to human volunteers.[19]

Operational advantages: these includes factors like cost and regulatory relief. Recent increases in the volume of confirmatory trials performed overseas have been driven in part

[†] Or, in the USA, persons who lack access to standard therapies because they are underinsured.

by reduced costs relating to labor, recruitment, and regulatory compliance.[20,21] The intensive technical requirements of gene-transfer studies tend to disfavor performing early-phase trials in LMICs, and indeed, one report suggests that such trials are disproportionately low in places like Eastern Europe, Latin America, and Asia.[20] Nevertheless, the placement of gene-transfer trials overseas has occasionally been motivated by cost and regulatory relief; Martin Cline's β-thalassemia experiment provides one such example. Rather than conduct his study at UCLA, Cline traveled to Italy and Israel.[‡] Though he persistently claimed otherwise,[22] the impression that Cline was motivated by regulatory relief is supported by the fact that he did not inform the Israeli authorities or patients of his intention to use recombinant DNA, and prevaricated when confronted by NIH authorities.[23] According to news reports, researchers in the USA and Europe are increasingly turning to China to pursue their gene-transfer studies; in the words of one USA-based researcher, "if I were making a long-term investment in biotech, and particularly in gene therapy, I would be making it in China, not here. They have figured out how to get [gene therapy] approved."[24] That China provides a "large number of highly cooperative patients" whose cancer treatments are not covered by the healthcare system[25] is an added bonus.

Justice and translational trials

Do each of these five rationales for recruiting volunteers from economically disadvantaged settings withstand ethical scrutiny?[§] Specifically, do they raise questions of fairness? Of the three principles said to ground the ethics of human experimentation, justice is by far the most under-attended. Many commentators have criticized research ethics for focusing too heavily on means rather than ends,[26,27] but discussions of justice force contemplation of ends. And of the three canonical principles of research ethics, justice is also likely to be the least familiar for clinicians. Physicians are fiduciaries; their training emphasizes that primary allegiances are owed to their patients, not others. Except in circumstances like triage, extreme scarcity, and public health, most would rightly condemn clinicians who ration expensive therapies or make other medical decisions by appeal to the needs of others.

Unlike care, however, research is pursued primarily to advance societal ends. Questions of morality thus arise if an investigator's activities promote societal ends at the expense of disadvantaged groups, or if one group's deprivations are used to further the moral interests of those who are better off. Indigent, institutionalized, or imprisoned patients have historically shouldered disproportionate burdens in medical research. The use of such dependent populations was occasionally defended as providing an opportunity for subjects to "return to the community for the care devoted to them."[¶] The National Commission's Belmont Report was the first document on research ethics to formally reject this line of reasoning:

---

‡ Of course, Italy and Israel cannot be properly identified as economically disadvantaged countries. Here, the disadvantage was largely political: neither had developed as formalized a system for ethical oversight of research as the USA.

§ Obviously, LMICs are not a uniform vista of poverty; nor are HIC populations all affluent. This chapter is not so much concerned with recruitment of affluent LMIC populations. Similarly, the arguments made here would apply equally to HIC-sponsored studies that recruit impoverished or disadvantaged HIC populations.

¶ See Lederer SE. *Subjected to Science.*[28] This argument is not as antiquated as it may appear. One researcher once told me (anonymously) "It cannot help the cause of the developing world that there are companies in these countries that specialize in producing knock-off drug products at marginal cost/price while ignoring intellectual property rights that are awarded in the developed world on the basis of

> ... selection of research subjects needs to be scrutinized in order to determine whether some classes (e.g. welfare patients, particular racial and ethnic minorities, or persons confined to institutions) are being systematically selected simply because of their easy availability, their compromised position, or their manipulability, rather than for reasons directly related to the problem being studied. Finally, whenever research supported by public funds leads to the development of therapeutic devices and procedures, justice demands both that this not provide advantages only to those who can afford them and that such research should not unduly involve persons from groups unlikely to be among the beneficiaries of subsequent applications of research.[29]

Not until the mid 1990s, however, was justice catapulted from a backwater for research ethics to its leading edge. The triggering event was a series of trials in Africa and Asia aimed at controlling mother-to-child transmission of HIV. Wealthy sponsor countries proposed to test whether a shortened course of AZT could reduce HIV transmission in resource-poor settings. Though a longer course of AZT (the ACTG 076 regimen, so named because it was the 76th protocol tested by the AIDS Clinical Trials Group) had been proven to cut transmission rates significantly, investigators proposed to test their shortened course against a placebo. Such withholding of standard care for a life-threatening disease would never have been tolerated in affluent countries. Critics, who included many prominent figures,[30] charged that the high ethical standards of trials in industrialized countries should be observed in the developing world. Defenders of similar authority[31] argued that an ACTG 076 control group would be next to impossible in resource-poor settings, that using placebos was not inconsistent with a local standard of care (which was nothing), and that halting such trials would severely impede the development of economically viable measures to control perinatal transmission of HIV in low-income countries.

The debate prompted a series of reports and revisions to ethics codes. The 2000 revisions of the Declaration of Helsinki, for example, incorporated several new provisions. The first stated "some research populations are vulnerable and need special protection. The particular needs of the economically and medically disadvantaged must be recognized." The second stated "[m]edical research is only justified if there is a reasonable likelihood that the populations in which the research is carried out stand to benefit from the results of the research." The third major addition stated "[a]t the conclusion of the study, every patient entered into the study should be assured of access to the best proven prophylactic, diagnostic and therapeutic methods identified by the study." In what follows, I will refer to the latter two standards as "responsiveness" and "post-trial access," respectively.**

Though specific language varies, both conditions were taken up by numerous other authoritative policy statements and reports, including CIOMS,†† the Nuffield Council, the National Bioethics Advisory Committee, the European Group on Ethics and Science of New

---

investment in research and development... if the developing world is to (eventually) derive benefit from technological advances made largely through the efforts or private R&D and government sponsored basic science research (e.g. NIH) in the developed world, why can't developing countries make a contribution by participation in the clinical trials?"

** Before proceeding, I should note a deficiency in various ethics codes with respect to their applicability to the examples described above. The language of most codes concerns trials conducted by high-income country sponsors in resource-poor settings. In contrast, many gene-transfer studies described in this chapter are conducted within the sponsor country. But whether subjects are imported or protocols exported, the ethical principles should not vary.

†† CIOMS Guideline 10 states that "Before undertaking research in a population or community with limited resources, the sponsor and the investigator must make every effort to ensure that: the research is responsive to the health needs and the priorities of the population or community in which it is to be carried out; and

Technologies to the European Commission, and South Africa.[32–36] Some policies, like India's[37] and Nepal's,[38] endorse the former but do not specifically address the latter in the context of studies involving economically disadvantaged populations.

---

any intervention or product developed, or knowledge generated, will be made reasonably available for the benefit of that population or community."[32]

The Nuffield Council states that "Researchers should endeavour to secure post-trial access to effective interventions for all the participants in a trial who could benefit. In addition, the possibility of introducing and maintaining a successful treatment in the wider community should be considered before research is conducted. If it is thought that this will not be possible, researchers must justify to the relevant research ethics committee why the research should be carried out." (xvi; 9.31); "we consider that all externally-sponsored research should be required to fall within the ambit of the national priorities for research related to healthcare within developing countries, unless the reason for not doing so can be justified to the appropriate research ethics committee within that country." Elsewhere, the report adds "We therefore recommend that when research funded by external sponsors is proposed which falls outside the national priorities for research into healthcare set by a host country, those proposing the research be required to justify the choice of the research topic to the appropriate research ethics committees in both the host and sponsoring countries" (2.32)[33]

The National Bioethics Advisory Committee's Recommendation 4.1 states "Researchers and sponsors in clinical trials should make reasonable, good faith efforts before the initiation of a trial to secure, at its conclusion, continued access for all participants to needed experimental interventions that have been proven effective for the participants. Although the details of the arrangements will depend on a number of factors (including but not limited to the results of a trial), research protocols should typically describe the duration, extent, and financing of such continued access. When no arrangements have been negotiated, the researcher should justify to the ethics review committee why this is the case." Recommendation 1.3: "Clinical trials conducted in developing countries should be limited to those studies that are responsive to the health needs of the host country."[34]

The European Group on Ethics in Science and New Technologies to the European Commission says:. "In industrialised countries, free supply of a proven beneficial new drug to all the participants of a trial after the trial is ended is the rule as long as it is not yet available through the normal health care system. In developing countries, the same rule must be applicable even if this implies supplying the drug for a lifetime if necessary. Moreover, there should be an obligation that the clinical trial benefits the community that contributed to the development of the drug. This can be e.g. to guarantee a supply of the drug at an affordable price for the community or under the form of capacity building." (2.13); "In the evaluation of a research protocol, special attention should be paid to the following issues: the relevance of the research to be carried out in a developing country. Specific attention should be paid when the objective of the clinical trial does not comply with health priorities of the host country" (2.9).[35]

The South African Department of Health says "Whilst maintaining the highest standards of clinical research it is important that clinical trials are based on priority, country specific research questions. Relevant and important questions should also be problems that significantly affect local and regional population. Study rationale should demonstrate that the study question has not been substantially answered and that adequate systematic review of the subject under discussion was done. The findings of which must be translatable into mechanisms for improving the health status of South Africans. Solutions should have the potential for implementation … A study on a population comprising a vulnerable group must be able to justify the choice of that population on scientific and social justice grounds. The researchers should be able to say why they have chosen that vulnerable group rather than the other and if a population that is a vulnerable group will stand to benefit in ways that reduce their vulnerability".[36]

The Indian Council of Medical Research says "persons who are economically or socially disadvantaged should not be used to benefit those who are better off than them".[37]

Nepal's guidelines state: "The following conditions have to be considered before external sponsors can undertake research among the Nepalese people…(a) The research is preferably responsive to the health needs and priorities of Nepal as well as being sensitive to the existing culture and social values; (b) The research cannot be carried out reasonably well in the sponsors' country." Elsewhere, the guidelines bar externally sponsored phase 1 studies: "In accordance with this principle, a new drug or appliance developed elsewhere can only be tested in the Nepalese population after a Phase 1 trial has been conducted elsewhere."[38]

Responsiveness, post-trial access and translational trials

Post-trial access has thus far received the most sustained attention. This requirement has variously been interpreted as applying either to trial subjects (who might relapse after completion of the trial if effective study medications are withdrawn) or to their communities. Whereas the Declaration of Helsinki mandates post-trial medication access for subjects, NBAC and the Nuffield Council view this as infeasible and merely encourage it. Two additional criticisms have been leveled against the post-trial access requirement. First, some argue that it is too restrictive: trial subjects or their communities can benefit in many ways from hosting a trial, and various commentators urge a requirement that would permit other benefits, such as medical training, employment, infrastructural investment, or financial compensation.[39] Regardless of whether one finds this argument persuasive (see Box 8.1), a second criticism is that the requirement is inapplicable to early-phase trials, because they are not intended to vindicate experimental interventions.[39] Post-trial access to study interventions would thus not appear a reliable policy for ensuring the fairness of trials involving economically disadvantaged volunteers.

What about responsiveness? As noted, this standard is almost universally endorsed, though with varying degrees of stringency. Indeed, even critics of the Declaration of Helsinki language would seem to support responsiveness in some guise. For example, proponents of the Fair Benefit approach state that "research should have social value: it should address a health problem of the developing country population."[40] Nevertheless, few scholars and policy-makers have stepped forward with a definition, defense, and elaboration of conditions for its fulfillment. Clearly, whether translational trials that recruit volunteers in resource-poor settings can fulfill canonical notions of justice in research ethics will hinge in part on how "responsiveness" is defined and specified.

A minimalist interpretation would require that the trial investigate an intervention against a disorder that is sufficiently common in the resource-poor setting where the research is proposed. This appears to be the position of the "Fair Benefits" advocates, who defended SmithKline Beecham's trial of a hepatitis A vaccine, Havrix®, in Thailand even though "it was unlikely that Havrix would be included in the foreseeable future in Thailand's national immunization program … [and] at the start of the trial, all collaborators recognized that the largest market for Havrix would be travelers from developed countries."[41]

This position, which uses biological criteria alone to measure responsiveness, is untenable. It is inconsistent with various research ethics policies that expressly assert social and political factors like "priority" and affordability in deciding whether a study is responsive. For example, South African guidelines state that "it is important that … trials are based on priority, country specific research questions … solutions should have the potential for implementation." Similar references to "priorities" in host countries are made by CIOMS and the European Group on Ethics and Science of New Technologies to the European Commission. And the Belmont Report highlights the importance of affordability.

Another reason to question a strictly biological definition of responsiveness is that it would bar only the most extreme types of study, those like the 1963 Jewish Chronic Disease Hospital, in which patients with one set of afflictions (chronic, debilitating disorders) were injected with a suspension of cancer cells to study another disease (cancer).[‡‡] Practically any trial involving any disease would be considered responsive so long as a single case was documented in the

[‡‡] There is an interesting link between this research and gene transfer: the main investigator in this study was Chester Southam, who pioneered oncolytic viral therapy in the 1950s.[42]

host community. Why would so many important committees have endorsed responsiveness if so little was intended?

More plausible is the suggestion that disease states – and interventions targeting them – are situated within a social context. Advanced healthcare technologies offer nothing to patients without a budget to procure them, an infrastructure to distribute them, and a team of clinicians to dispense them. Similarly, illnesses in the developing world are often as related to material deprivations (for example, ability to purchase medications) as they are to biological states.

Responsiveness, then, should recognize the social, political, and economic conditions necessary for actualizing a given intervention's health benefits. Alex London and I have argued that for a study to be responsive to a resource-poor community, it should fulfill two conditions:

Urgency: it must "address an 'urgent'[§§] health need of the host community." Whether a health need is 'urgent' will depend on the severity of the condition and on its prevalence in the host community, or among special subpopulations, such as children, that play an important role in disease transmission (further elaboration of urgency will be provided later). We do not consider a health need "urgent" if it can be addressed more effectively or efficiently through the mobilization of available resources.

Local utility of knowledge: a protocol should aim to expand the capacity of health-related social structures in that community to meet urgent health needs. Fulfilling this second criterion requires either that the protocol aim at gathering knowledge to enable hospitals, clinics, or public health entities to better address the health needs of community members, or that the protocol be part of a process of developing an intervention that can reasonably be projected to be deployed in the host population.[43] One measure of whether an intervention is deployable in LMIC communities is whether it can be provided in a variety of settings by personnel who might not have received extensive training (so-called "close to the client" health systems).[44]

Box 8.1 Can "Fair Benefits" assure justice? Appraising an alternative to responsiveness

Proponents of the "Fair Benefits" position defend a very permissive definition of responsiveness; they argue that fairness is about "how much" host communities benefit rather than "how" they benefit.[39] To "Fair Benefits" proponents, requirements like reasonable availability and responsiveness are restrictive and perhaps even paternalistic, because they prescribe "how" communities should benefit.

The "Fair Benefits" view was partly developed to respond to the problem of phase 1 studies in LMICs: were reasonable availability and responsiveness the appropriate ethical standards, host communities would be deprived of any benefits after phase 1 trials – especially where they fail to justify advancing to later stages of development. Instead, the "Fair Benefits" view is that host communities should be free to negotiate a package of benefits with sponsors however they wish. Rather than ask for access to study medications, they might instead negotiate training opportunities for LMIC healthcare workers, equipment and facilities, or medicines unrelated to the study.

The "Fair Benefits" position has a number of attractive features, including its respect for host community autonomy and the fact that it is ostensibly more compatible than "reasonable availability" with respect to nonconfirmatory clinical trials like phase 1 studies. So why not simply dispense with "responsiveness" and go with "Fair Benefits?"

I believe that the "Fair Benefits" view – at least in its present version – is flawed. For one, it advances a "bargaining" model for pre-trial agreements. This is problematic in three important

[§§] One set of commentators suggested "priority" instead of "urgency." The former would seem compatible with the way that London and I intended "urgency;" as such, I will use the two interchangeably.

respects. It assumes bargaining parity.[45] The reality, however, is that where sponsors can choose among multiple trial locations, host communities will tend not to have the luxury of choosing sponsors or studies. Bargaining will thus tend to favor the HIC sponsor. It also does not recognize informational asymmetries: while drug companies will keep secret their projected revenues for a particular product line, projected benefit to host communities is more transparent. As a consequence, sponsors are in a better position to calculate the least they can offer before host communities will walk away from the bargain; host communities do not have the requisite information to determine the most they can demand before HIC sponsors walk away. The bargaining model assumes that those bidding on behalf of host communities legitimately and adequately represent their communities, and that negotiated benefits will be directed in an equitable way. This might not be a prudent approach in many LMICs, where oppression of ethnic minorities is not uncommon, as is corruption among government officials and their healthcare ministries.

Another flaw in the "Fair Benefits" view is the suggestion that "how much" counts, not "what." But different categories of benefit have properties that affect who, how many, and how much different parties benefit. Surely these become relevant in evaluating whether a transaction is just. One type of benefit posited by the "Fair Benefits" proponents, medical training for LMIC healthcare workers, helps make this point. People walk, and they take their training with them. As a consequence, one of the major challenges confronting LMICs is that healthcare personnel with the most advanced training tend to migrate to HICs, leaving LMICs bereft of direct benefits (apart from remittances). Knowledge benefits provide a contrast. Unlike, say, a refrigerator, knowledge is a non-rival good, and its marginal costs of production are almost zero. Whereas using a refrigerator prevents others from using it, one person's use of knowledge does not affect whether another uses it, and knowledge gains can be distributed without major expense. Knowledge benefits of the sort promoted by responsiveness thus tend to favor broader populations.

There is one important respect where Fair Benefits proponents have a point: however universal the endorsement of responsiveness, its philosophical foundations are not well worked out. Typically, proponents of responsiveness defend the concept by appealing to risks and benefits. The language of the National Bioethics Advisory Committee report on international research exemplifies this: "In the research context, distributive justice demands that no one group or class of persons assumes the risks and inconveniences of research if that group or class is unlikely to benefit from the fruits of that research."[46] I believe this emphasis on risk is misguided, because there are many instances where risk–benefit balances are highly favorable to individuals and host communities, but where fairness issues linger. Imagine if the short-course AZT studies that instigated debates about these requirements were mainly run to help insurance companies in HICs lower the costs associated with AIDS care. Though African mothers receiving placebo would have been denied a possibly effective intervention, they and their communities arguably would have benefited from better monitoring and counseling while enduring few if any new risks or burdens. The Milan ADA-SCID study is perhaps a clearer case: because volunteers were imported into HIC settings, LMIC communities did not bear any of the costs or burdens of supporting the study.

Any defense of responsiveness must address directly the problems raised when advantaged communities intentionally derive benefits from another group's unfair disadvantages, and why this is morally even more troubling in the context of the production of medical knowledge. Nevertheless, this chapter does not attempt a principled defense of this position, and accepts the view that the concept draws its moral force from the fact that so many august, culturally and economically diverse, and deliberative bodies have consistently endorsed it.

Presuming responsiveness?

Could the various types of gene-transfer studies described above be deemed to fulfill these conditions? Some gene-transfer researchers argue that early-phase gene-transfer studies like those described above should be presumed responsive to LMIC communities.[¶¶] They support their reasoning with four arguments: "properties," "autonomy," "polyvalence," and "trickle-down." The "properties" argument runs as follows: stable, one-time interventions for diseases like hemophilia would eliminate the need for chronic and costly treatment regimens. Cancer gene-transfer products are easier to administer than standard anti-cancer drugs, and because they have a diminished toxicity, patients who receive them are easier to manage in resource-constrained settings.[48] The "autonomy" argument notes that researchers and many volunteers in countries like Brazil welcome such studies; LMIC interest in gene transfer is illustrated by the fact that China ran the first hemophilia gene-transfer trial and is the first country to commercialize two cancer gene-transfer products, and that middle-income countries like Mexico,[49,50] Brazil,[51] the Philippines,[52] and Malaysia[22,53] have active gene- and cell-transfer research programs. For HIC ethicists like myself to question the responsiveness of these trials is at best misguided, and at worst a form of "ethical imperialism." According to the "polyvalence" view,[6] results from phase 1 trials of novel interventions often inform diverse research programs. This is consistent with what I argued in Chapter 6. Success in ADA-SCID studies has important implications for other inherited hematopoietic diseases,[54] some of which (e.g. thalassemia and sickle cell anemia) are more prevalent in LMICs. A fourth argument ("trickle-down") concedes that some gene-transfer interventions will be taken up more rapidly in HICs. But because such treatments will eventually be taken up in LMICs, gene-transfer trials recruiting volunteers in LMICs can be considered responsive.

With respect to the properties of gene-transfer agents, it is certainly true that many gene-transfer interventions have properties that would make them amenable to addressing priority health issues in LMICs. It also seems reasonable to suggest that global health calamities like malaria, tuberculosis, and HIV/AIDS will not be tackled effectively without the use of cutting-edge technologies like those involving gene transfer. Nevertheless, such arguments do not automatically confer the quality of responsiveness on gene-transfer studies. First, the criterion of urgency might be difficult to satisfy in certain circumstances (see Box 8.2). Second, whether knowledge or interventions are deployable will also be in question. As indicated above, technologies require certain social and infrastructural inputs. Just as refrigerators cannot keep food from spoiling unless there is a grid to supply electricity, drugs cannot save lives unless they can be deployed to patients who need them.

Medical skill and facilities will often limit the ability of LMICs to deploy interventions. The Milan ADA-SCID study, for example, involved isolating the CD34+ cells of volunteers, genetically modifying them, and infusing after four days. In the intervening time, subjects were carefully monitored as they received nonmyeloablative conditioning – a treatment that uses the chemotherapy drug busulfan to kill off cells in the bone marrow. Performing this procedure in LMICs would require expensive equipment and adequately trained medical personnel.

Healthcare budgets in LMICs will also strain access to gene-transfer products. Drugs used to treat rare "orphan" disorders are notoriously expensive: the yearly cost of enzyme (or factor) replacement for ADA-SCID, Gaucher's disease (one of the more common lysosomal storage diseases), and severe hemophilia are roughly \$200 000,[4] \$170 000,[55] and \$60–150 000[56], respectively. Moreover, pharmaceutical companies typically price new drugs two to three times higher than existing products if they offer significant therapeutic advantages.[57] It seems

¶¶ This position is often tacit, especially in the literature on hemophilia gene transfer. An explicit expression can be found in Ponder & Srivastava.[47]

highly improbable that companies manufacturing newer-generation orphan drugs will substantially lower prices, or that LMICs will be able to absorb such costs.

Experience thus far highlights some of the impediments that are likely to be encountered in attempting to deploy gene-transfer interventions in LMICs. For example, news reports indicate that Avigen had intended to price the AAV-based hemophilia treatment described above at $400 000 for each administration (occurring once every several years).[58]*** While this might represent a cost saving over current treatment regimes, it seems unlikely that it will dramatically improve the economics of hemophilia care in LMICs. In China, a typical course of treatment with Gendicine® reportedly costs over $3000 – a figure that exceeds per capita yearly income in the country's wealthiest city, Beijing (approximately $2700). The drug is also much more expensive than alternative treatments like radiotherapy. At present, Gendicine is not covered by any health insurers, and most patients receiving it are medical tourists from HICs who are unable to obtain the medicine in their own country.[60,61]

Box 8.2 Are rare diseases "urgent"?

Some might question whether trials involving rare diseases could ever be considered urgent, given their low prevalence. Certainly, disorders like lysosomal storage diseases, brain cancer, and primary immunodeficiencies are serious and are indeed urgent in the sense that they demand medical attention. However, in most low-income countries and many middle-income countries, such diseases are unlikely to represent priority health problems, because the purchase of medications to treat them would likely divert resources from medicines that could benefit larger numbers with similarly urgent needs.[62]

In recent years, some rare-disease advocates have advanced strong arguments consistent with a view that rare diseases can be urgent. These advocates point out that excluding persons with rare disorders from healthcare services is unjust,[63] and note that the World Health Organization includes hemophilia treatments on their Essential Medicines List (EML). Because the World Health Organization defines essential medicines as "those that satisfy the priority health care needs of the population,"[64] advocates would appear to have a reasonable claim that global consensus favors classifying rare diseases like hemophilia as a priority disease for LMIC healthcare systems.

Advocates are doubtless correct in pointing out the injustice of excluding persons from healthcare systems simply because their afflictions are rare. However, the question here is not whether such persons are excluded from care, but rather, whether their illness should represent a public health priority. Wherever resources for health care are limited, tragic choices will be inevitable. Arguably, an even greater injustice occurs if many persons with life-threatening illnesses are unable to access medicines because scarce resources have been allocated to a smaller group of patients.[62] Moreover, the suggestion that the EML's listing of hemophilia drugs is evidence of a global consensus on the priority of rare diseases is not altogether persuasive. For one, factor concentrates represent the only treatment for a Mendelian genetic disorder on the EML. Second, the expert committee that recommended the inclusion of hemophilia noted the "inherent inconsistency" of this decision.[65] Last, hemophilia would seem to be a "special case" in that failure to treat it effectively can have broader public-health implications, given the potential for spreading blood-borne diseases like hepatitis and HIV.

*** To put these figures in perspective, consider the example of Brazil. According to a 2003 report, the Brazilian government allocated $11 000/year for the care of each patient with hemophilia. The country provides factor treatments free to all diagnosed patients. Unlike in the USA, however, primary prophylaxis and immune tolerance induction are not provided, and recombinant factor is not routinely used.[59] Thus, though many Brazilians have access to quality care for hemophilia, the nature, extent, and expenditures on that care are significantly behind those of countries like Canada and the USA.

The "autonomy" argument uses the enthusiasm of host community researchers and study subjects as evidence that gene-transfer studies are "responsive." Certainly, researchers and individual patients will often have knowledge of "on the ground" realities that might affect whether trials are responsive to a host community. Nevertheless, sizing a study's responsiveness requires an assessment of whether an intervention addresses an urgent host community health need, and whether it is deployable within the community. The former is, in part, a social and political judgment best left to policy-makers; the latter involves knowledge of economics, sociology, and healthcare administration that researchers (or volunteers) will often lack.

The "polyvalence" argument would posit all gene-transfer trials as responsive because phase 1 trials can produce insights that can lead to applications different than those contemplated by the original study. Though true, it seems unwise to justify ethically contentious actions by appealing to remote and unplanned outcomes. Even if, as I have argued in earlier chapters, iterative and reciprocal values are to be integrated into the planning of translational trials, sponsors will still need to provide a coherent story of how such findings would be assimilated to clinical practice within a reasonable time-frame.

The "trickle-down" view is that, since gene-transfer interventions developed primarily for HIC populations will eventually reach LMICs, any gene-transfer study should be deemed "responsive." This argument has a number of problems. First, it contradicts outright the notion of responsiveness. To be responsive is to assess the particularities of a given situation, and to adjust one's actions accordingly. Passive processes like "trickle-down" are antithetical to assessment and adjustment. Second, as suggested previously, subscribing to this view would make nearly any conceivable trial responsive. If this were the case, why would so many organizations, beginning with the National Commission and continuing through to the World Medical Association, have gone to lengths to articulate statements requiring responsiveness? Third, this position would seem to consider the current rate of technological diffusion to be an acceptable standard against which to measure the impact of efforts to improve the health of LMIC populations. Yet consider the pace at which interventions are translated to LMIC contexts. As much as 90 percent of avoidable mortality in LMICs is attributable to diseases for which effective social or medical interventions already exist.[44] Closer to the example of gene transfer, per capita factor VIII use for treating hemophilia in the two countries where Avigen recruited volunteers for its hemophilia study, Brazil and India, is approximately 36 percent and < 1 percent the level used in G7 countries, respectively.[66] Only the most hardened cynic would defend the current rate of technological diffusion to LMICs as acceptable.

Nevertheless, the trickle-down argument raises important questions. Because phase 1 studies are not confirmatory, deployment of interventions will always be delayed to some extent. And because the largest markets for most gene-transfer products will be in HICs, it seems reasonable to predict that products will be marketed in HICs before they are launched in LMICs. Surely, some anticipated delay between trials and deployment is compatible with responsiveness. Precisely what degree, however, is a question I leave to others.

Now let's examine some of the motivations driving the recruitment of LMIC populations into HIC-sponsored studies. Some types of study will face tough sledding in trying to claim they are responsive in the way it was defined above. For example, the category of "fortuitous studies" recruits volunteers precisely because they are unable to access therapies that are standard in HICs. The very fact of their restricted access to leading therapies would seem to make them presumptively *unlikely* to access new leading therapies. Similarly, studies that recruit in LMICs for reasons of expedience will also tend to have a more difficult time claiming responsiveness. In the hemophilia study described above, patients in HICs can adequately – if not optimally – control their disease using recombinant coagulant factor. A phase 1 trial involving an intervention that is not well characterized with respect to risk will tend to be unattractive for HIC volunteers, especially if it requires (as do most hemophilia protocols) subjects to forgo standard therapies during the trial. In contrast, people who are unable to access safe and effective factor-replacement therapies will tend to view the study more favorably. In this case, at least, expediency would seem to be "fortuity in disguise."[43] Studies placed overseas to escape HIC regulation will also tend[†††] to face difficulties claiming responsiveness. Such studies make no pretense of being part of a process of developing applications that are deployable in LMICs. Instead, they allow sponsors to reach HIC markets without having to comply with certain cumbersome HIC regulations during development. What makes such studies particularly troublesome is that they shift risks deemed unacceptable by HIC regulators to jurisdictions where the knowledge benefits from a completed study are significantly less likely to reach local populations.

Studies involving ultra-rare diseases will in many cases be impossible without expanding recruitment to LMICs. Such studies raise less acute concerns about responsiveness because they are seeking LMIC volunteers in spite of disadvantage, rather than because of it. Prevalence is probably the easiest to reconcile with responsiveness, because such studies are aimed squarely at tackling diseases that primarily affect persons in LMICs. Nevertheless, neither ultra-rarity nor prevalence establishes that an intervention is deployable within LMIC communities. For example, HIC sponsors often undertake research in tropical diseases because they affect military deployments or tourists. If products emerging from such studies seem unlikely to be used among the LMIC populations in which they were tested, such studies fail responsiveness.

Reconciling early-phase gene-transfer trials with responsiveness

The logic of the foregoing would appear to write off certain categories of studies, like the Milan SCID study, as unethical. But note that the definition of responsiveness provided above appeals to statements of fact: does an intervention address an "urgent" health need, and is it deployable? I propose a two-step process by which sponsors and researchers might make the case that their studies – even ones driven by fortuity or expedience – might fulfill responsiveness.

The first step involves marshaling evidence to rebut a baseline presumption that a study is not responsive. Some researchers will no doubt find a presumption against responsiveness to be extreme. But consider how ethics committees and policy-makers function: as noted in Chapter 5, new drugs are presumed to be risky and ineffective until proven otherwise. By law, drug companies are required to submit evidence to regulatory authorities to rebut this presumption. Similarly, IRBs and DSMBs are designed to provide a skeptical vetting of research protocols by examining the quality of supporting evidence, the qualifications of investigators, the validity of study design, and the comprehensiveness of disclosure during informed consent. Building the case that a study is responsive should involve a similar process in which the research team persuades independent and critical oversight bodies by using evidence and inference.

Consistent with the definition of responsiveness, this evidence should speak to the two components of responsiveness described above (see Table 8.1). First, is a particular health condition considered urgent in the host community? Answering this question might involve

[†††] I say "tend," because there are some circumstances where regulatory variations reflect genuine moral disagreement rather than underdevelopment. One example is the use of embryo-derived tissues.

Table 8.1 Evidentiary requirements for fulfilling responsiveness

Condition	Factors warranting evidence
Is the health condition "urgent" in host communities?	Policy priority of disease
	Disease prevalence
	Disease morbidity or mortality
	Social or economic impact of disease
	Effect on special groups
Can the most likely intervention be foreseeably deployed in host communities?	Projected costs of most likely intervention
	Per capita healthcare expenditures
	Cultural barriers to uptake for most likely intervention
	Plans for distribution of intervention
	Infrastructure and training requirements for intervention
	Intention of sponsor to seek timely licensure of product

showing that a condition has been identified by policy-makers as a priority; that a disease is prevalent; that the disease is highly morbid or lethal; that the condition imposes particularly large social or economic impacts on host communities; that it afflicts groups that deserve special protection (e.g. children); or that it could transmit the disease to wider populations (e.g. sex workers).

The second step is to ask whether the intervention – with only minimal alteration – is capable of deployment in host communities. Answering this question will require addressing projected costs of the intervention and healthcare budget impacts; per capita health care expenditures; whether treatment costs are borne by third parties or individuals; whether there might be significant cultural barriers to applying the intervention; demands that might be placed on equipment or healthcare personnel; infrastructural requirements; plans for distribution of the intervention; and perhaps above all, whether commercial sponsors are on public record with plans to seek timely licensure in host communities.[‡‡‡]

If answers to the above questions are satisfactory, the trial is responsive. I suspect that many vaccine studies will have little difficulty meeting an acceptable evidentiary threshold of responsiveness. But other studies – like those involving fortuity, expedience, or regulatory relief – will find the process more onerous. These would then proceed to the next question: do sponsors have a compelling and publicly stated plan to overcome access barriers identified in the first step?[§§§]

[‡‡‡] Some pharmaceutical products available in HICs are never actually launched in LMICs. Others are launched with significant delay. For example, Gilead first started selling in HICs its anti-retroviral product Tenofovir disoproxil in October 2002. Almost three years later, the company had filed materials for registration in only 11 of 53 African countries. See Ford & Darder.[67] According to one study, the median delay between a drug's first country launch and its availability to US consumers is 20 months; the corresponding figure for LMICs is approximately 50 months. See Lanjouw.[68]

A matrix developed by the Bill and Melinda Gates Foundation-funded Diseases of the Most Impoverished Program might further inform this list of evidence items. See Clemens & Jodar.[69]

[§§§] Some might find the provision of this second step too permissive in that it allows HICs to conduct trials in resource-poor settings with the primary objective not of advancing the interests of persons in the developing world, but rather, of persons in high-income countries. These critics are unlikely to find

Table 8.2 Policy options for overcoming responsiveness barriers

Mechanism	Advantages	Disadvantages
Pharmaco-philanthropy	No cost to LMIC	Dependence of LMIC on HIC charity
		Potential for use in marketing and intellectual property enforcement
Discriminatory pricing	Can be in the economic interest of HIC manufacturers	Costs might still be too high for some LMICs
	Taps in to the power of markets	Costs to HICs (e.g. leakage, political pressure in HICs for reduced prices)
Equitable access licensing	Suited to early stages of product development	Generic manufacturers might nevertheless be unwilling to produce product
	Uses patents to tap power of markets	
Advance-purchase agreement	No cost to LMIC for intervention access	Potential for diversion of charitable and host country resources in LMIC
	Provides market incentive to manufacturers	

Policies for achieving responsiveness

At least four non-exclusive policy options might enable sponsors to answer the second question satisfactorily (see Table 8.2). All relate to mechanisms for altering the economics of supplying interventions. When there is a firm, prospective commitment to such mechanisms, concerns about urgency and deployment relax – the former because products are made available through channels that are less likely to compete with a country's other health priorities, and the latter because sponsors themselves undertake or facilitate the deployment of an intervention in host communities.

The first supply mechanism would be a "pharmaco-philanthropy" program that would provide access to interventions in the event they prove efficacious in subsequent studies. Though probably the least optimal of the four in that persons in LMICs are dependent on the charity of HIC sponsors, pharmaco-philanthropy might nevertheless be appropriate in certain situations. Two programs provide an indication of how this should and should not be done. The first is Novartis's Glivec® International Patient Assistance Program (GIPAP), which provides imatinib mesylate (also known in North America by the tradename Gleevec®) to patients with chronic myelogenous leukemia in the developing world. According to reports, the program has provided free treatments to 14 500 persons worldwide who are uninsured and unable to purchase the drug themselves; countries that participate in GIPAP range from

trials like the Milan ADA-SCID ethically defensible under any circumstances. I offer two responses. First, what is argued here is intended to provide a modest and pragmatic proposal for reconciling the self-directed interests of research sponsors with a baseline standard for ethical conduct. Second, a position that mandates that LMICs always be the primary motivation for HIC sponsors is at odds with current systems in drug development. Drug development is primarily driven by profit rather than humanitarian considerations. Perhaps it should be different, and drug companies should be nationalized. Certainly, governments could do much more with tax and procurement policies to incentivize greater private sector investment in LMIC health needs. But for now, removing profit incentives by compelling drug companies to prioritize LMIC health needs when locating trials overseas would staunch drug development directed towards LMIC needs.

Albania to Zimbabwe. GIPAP works as follows: physicians in developing countries submit applications for their patients. The USA-based Max Foundation, which administers GIPAP, assesses applications on the basis of medical need, physician expertise, and monitoring capacity. Imatinib mesylate is then dispensed to eligible patients though the country's healthcare system.[70]

GIPAP does not appear to be an acceptable model in that the program seems less a foreign aid mechanism than a business instrument. Countries are only eligible for the program provided that they confer strict intellectual property protections for imatinib,[70] and Novartis abruptly cancelled India's GIPAP program when the country licensed generic Gleevec.[71] Novartis also appears to have recruited GIPAP beneficiaries into campaigns aimed at pressuring national healthcare systems to reimburse drug purchases or refuse licensure for generic versions. Indeed, an internal Novartis memo from the president of Novartis's oncology unit described the program as "achiev[ing] therapeutic and business goals" for the company, and an internal slide presentation described its objectives as having the aim of "OK Reimbursement."[71]

Genzyme seems to have implemented a more attractive model of pharmaco-philanthropy. Their example is also directly relevant to the current discussion in that the company has an active gene-transfer program and acquired Avigen's hemophilia technologies in 2005 (as noted above, it also has a record of recruiting subjects from LMICs).[72] Through Project Hope, Genzyme dispenses its enzyme replacement product, Cerezyme, to LMIC patients unable to afford the Gaucher's disease treatment.[73] As of 2006, 213 patients in 16 countries were receiving the drug.[72] To this author's knowledge, Genzyme has not come under any criticism for leveraging their charity arm for commercial gain – though there are many who are critical of the company's pricing and marketing practices.[74–77]

The second mechanism is what economists call discriminatory pricing (also sometimes called tiered-pricing), in which drugs are priced substantially lower in LMICs than in HICs. In some instances, this will actually be in the best economic interest of the drug manufacturer, since larger markets can be reached.[¶¶¶] Vaccine manufacturers have often used discriminatory pricing,[81] and large buyers of vaccines like UNICEF often condition their purchase on a discriminatory pricing regime.[80] One specific example where discriminatory pricing has been used is Novartis's anti-malarial drug Coartem, for which an adult course costs $40 in the USA, but $12 in malaria-prone regions and $2.4 through the World Health Organization.[82]

For anti-retrovirals, discriminatory pricing has a mixed record in improving access, in part because even reduced prices are in some instances well above what people in LMICs can afford.[82] Key to fulfilling responsiveness using discriminatory pricing will be evidence that host community pricing is compatible with what persons in those countries – or their governments – are willing to pay for such products. Also key is developing mechanisms to counter some of the concerns that drug companies have when they use discriminatory pricing. One is "leakage:" that is, resale in wealthier countries of products purchased in low-price markets. For gene-transfer products that require specialized facilities and personnel, leakage might prove more manageable than for small-molecule drugs.

A third set of mechanisms involves various intellectual property licensing arrangements. Owners of patents relating to new products could license their intellectual property to generic manufacturers either in the LMIC or in HICs for export to LMICs. For instance, under

¶¶¶ Assuming the costs of marketing the agent are lower than the price, that there is no importation of lower-priced drugs into high-income countries, and that knowledge about discriminatory pricing does not undermine the drug firm's efforts to negotiate prices with consumers in HICs.[78–80]

"equitable access licensing," universities holding intellectual property on interventions would include provisions for international access in any licensing agreements with drug manufacturers. This approach is particularly promising for gene-transfer and other novel intervention approaches, because key patents often originate in universities. In addition, equitable access licenses would likely be established before a phase 1 study rather than as a result of it.

Yale University used equitable access licensing with its patent on the AIDS drug stavudine. It involved arranging an agreement with its licensee, Bristol Myers Squibb, to permit the sale of generics in South Africa.[83] Nevertheless, such licensing practices are relatively new and their effectiveness in securing LMIC access remains to be established. One clear limitation of such an approach for gene-transfer products is that it depends on the willingness of generic companies to undertake manufacture of a product. Diseases that have low prevalence, and products that have complicated manufacturing requirements, might tend to be less attractive for generic manufacturers.

One last policy alternative is the advance-purchase agreement. This would legally bind donors to purchase a product once it becomes available, and to distribute it in LMICs.[84] Thus, for example, entities like the World Health Organization, G7 countries, the World Federation of Haemophilia, or host country governments might, at the point of initiating a phase 1 trial, commit to purchasing a certain volume of a new gene-transfer product at a set price. This would have several salutary effects: first, it would provide an incentive for companies to persevere through product development. Second, it would provide a mechanism for making a product available to persons in resource-poor settings. Last, it would spur companies to invest more resources in production capacity for eventual products.

The advance-purchase agreement is not without critics. For instance, in committing resources towards a drug product, host countries or donors might divert their energies from more effective public health measures.[85] Nevertheless, this approach might be particularly promising for ultra-rare disorders, where relatively well-financed charities in HICs – like the one that funded the Milan ADA-SCID study – might commit to financing the delivery of a certain volume of an intervention to LMICs.

Such measures might be "mixed and matched" to diseases, products, and/or host countries. For example, advance-purchase agreements might be coupled with discriminatory pricing for products aimed at conditions such as hepatitis C that are relatively low in prevalence in HICs. I leave further development of policy options to others more expert in the economics; the point here is that there are a number of reasonably well-developed arrangements aimed at improving LMIC access to the fruits of medical research. All could be worked out in advance of a trial. Sponsors and investigators wishing to pursue studies in LMICs that are, on their face, non-responsive should be asked to demonstrate how they intend to overcome economic, social, and political barriers to make their study responsive.

Closing thoughts

This chapter leaves a number of loose strands that require follow-up work: how much delay between an intervention's testing and its predicted application should be allowed before a study should be deemed "non-responsive"? And how should policy-makers define the borders of a host community: by geography (and if so, at what level – city, country, or region?) or by class (and if so, which one – economic, cultural, or medical?).

There are a number of other issues that warrant attention as well. For example, what are we to make of the ethics of "home-grown" gene-transfer programs in LMICs, specifically, ones that are aimed less at serving local health than economic needs by catering to wealthy foreigners? A related set of issues concerns ethics and oversight of medical tourism involving non-validated interventions: are clinicians in LMICs who offer these services acting unethically? Are clinicians in HICs who travel to LMICs to offer these services to HIC clients acting unethically?

The analysis provided in this chapter centers on how researchers and sponsors might design LMIC studies that fulfill international standards of justice. I argued that, consistent with international statements on research ethics, early-phase trials conducted in LMICs should be "responsive" to host communities, and that responsiveness consists of pursuing research that enhances the capacity of local systems to address urgent health needs. When studies, on their face, seem non-responsive, researchers and sponsors should establish measures or advance agreements to ensure that host communities can meaningfully access the trial's knowledge benefits.

As I bring this chapter to a close, however, a broader question still looms: to what extent should medical researchers commit themselves to addressing LMICs' health needs not as a condition for getting preferred protocols approved, but instead as a primary goal? Current biomedical research is grossly skewed against the suffering of persons in LMICs; one often-cited statistic is that 90 percent of the global burden of disease receives only 10 percent of research funding. Some commentators make a persuasive case that researchers have affirmative, role-related obligations to direct their research towards pressing social needs, which include the health deprivations of poor populations.[86]

If so, there are two important ways that gene-transfer researchers can make distinctive contributions to global health. The first is by working towards deployable vaccines against LMIC afflictions like HIV/AIDS, malaria, and tuberculosis. The second involves a less radical departure from what most researchers are already doing. Many middle-income countries and, to a lesser extent, low-income countries, are currently undergoing an epidemiological transition in which chronic conditions like heart disease and cancer are replacing infection and nutritional deficiencies as primary drivers of mortality.[87] Treatments that are standard in HICs – surgery, intensive chemotherapy regimes, and therapies personalized to a patient's disease profile – have little to offer countries with limited healthcare budgets and human resources. Translational scientists should devote part of their research portfolio towards chronic disease treatments that might be applied in LMICs.

Innovative strategies like gene transfer can and will likely make essential contributions to global health (as noted, gene transfer is already working towards that end with HIV vaccine research). Exacting grudging concessions from HIC companies and investigators primarily interested in serving affluent markets will only go so far. The real imperative is making global responsiveness a general priority for the field and its sponsors.

References

1. London AJ, Kimmelman J. Justice in translation: from bench to bedside in the developing world. *Lancet* 2008; **372**(9632): 82–5.
2. Kalman L, Lindegren ML, Kobrynski L, Vogt R, Hannon H, Howard JT, *et al.* Mutations in genes required for T-cell development: IL7R, CD45, IL2RG, JAK3, RAG1, RAG2, ARTEMIS, and ADA and severe combined immunodeficiency: HuGE review. *Genet Med* 2004; **6**(1): 16–26.
3. Lyon J, Gorner P. *Altered Fates: Gene Therapy and the Retooling of Human Life*. New York, W. W. Norton & Company, Inc., 1996; 110.
4. Chan B, Wara D, Bastian J, Hershfield MS, Bohnsack J, Azen CG, *et al.* Long-term efficacy of enzyme replacement therapy for adenosine deaminase (ADA)-deficient severe combined immunodeficiency (SCID). *Clin Immunol* 2005; **117**(2): 133–43.
5. Aiuti A, Slavin S, Aker M, Ficara F, Deola S, Mortellaro A, *et al.* Correction of ADA-SCID by stem cell gene therapy combined

with nonmyeloablative conditioning. *Science* 2002; **296**(5577): 2410–13.

6. Kimmelman J. Clinical trials and SCID row: the ethics of phase 1 trials in the developing world. *Dev World Bioeth* 2007; **7**(3): 128–35.
7. Culver KW. The potential for genetic healing. *JAMA* 1992; **268**(13): 1768.
8. Manno CS, Pierce GF, Arruda VR, Glader B, Ragni M, Rasko JJ, *et al.* Successful transduction of liver in hemophilia by AAV-Factor IX and limitations imposed by the host immune response. *Nat Med* 2006; **12**(3): 342–7.
9. National Institutes of Health. *Safety of a New Type of Treatment Called Gene Transfer for the Treatment of Severe Hemophilia B.* In: ClinicalTrials.gov, ed. *Identifier: NCT00076557.* http://clinicaltrials.gov/ct2/show/NCT00076557?cond=%22Hemophilia+B%22&rank=10; 2004.
10. National Institutes of Health. *Gene Transfer for Subjects with Hemophilia B Factor IX Deficiency.* In: ClinicalTrials.gov, ed. *Identifier: NCT00515710.* http://clinicaltrials.gov/ct2/show/NCT00515710?term=hemophilia&phase=0&rank=2; 2008.
11. Green MA. The myozyme miracle: The team behind the treatment. *Inside Duke University Medical Center & Health System* 2006; **15**(16).
12. Sly WS, Vogler C. Brain-directed gene therapy for lysosomal storage disease: going well beyond the blood–brain barrier. *Proc Natl Acad Sci USA* 2002; **99**(9): 5760–2.
13. Meikle PJ, Hopwood JJ, Clague AE, Carey WF. Prevalence of lysosomal storage disorders. *JAMA* 1999; **281**(3): 249–54.
14. Testimony of Mark Sands, Recombinant DNA Advisory Committee (RAC). Minutes of Meeting June 21, 2006. MPS Type VII. Imports 3 subjects from Brazil (at p. 15).
15. National Institute of Allergy and Infectious Diseases. *NIAID Launches First Phase II Trial of a "Global" HIV/AIDS Vaccine.* US Department of Health and Human Services, National Institutes of Health, 2005.
16. National Institute of Allergy and Infectious Diseases. *NIAID Begins Enrolling Volunteers for Novel HIV Vaccine Study.* US Department of Health and Human Services, National Institutes of Health, 2005.
17. Seiler S. Targeted Genetics, IAVI and Collaborators Expand Phase I Clinical Trial Program of HIV/AIDS Vaccine Program to India. Targeted Genetics, *Press Release*, 2005.
18. Cohen J. AIDS vaccine research. Thumbs down on expensive, hotly debated trial of NIH AIDS vaccine. *Science* 2008; **321**(5888): 472.
19. Recombinant DNA Advisory Committee (RAC). Minutes of Meeting June 8, 2004. Update on Protocol #0107-488: A Phase I, Open-Label Clinical Trial of the Safety and Tolerability of Single Escalating Doses of Autologous CD4 T Cells Transduced with VRX496 4 in HIV-Positive Subjects 5.
20. Thiers FA, Sinskey AJ, Berndt ER. Trends in the globalization of clinical trials. *Nat Rev Drug Disc* 2008; **7**: 13–14.
21. Tufts Center for the Study of Drug Development. *Outlook 2007*; http://csdd.tufts.edu/InfoServices/OutlookPDFs/Outlook2007.pdf.
22. Lyon J, Gorner P. *Altered Fates: Gene Therapy and the Retooling of Human Life.* New York, W. W. Norton & Company, Inc., 1996.
23. Cook-Deegan RM. *Cloning Human Beings: Do Research Moratoria Work?* National Bioethics Advisory Commission Publications, 1997.
24. Einhorn B, Carey J, Hall K. A cancer treatment you can't get here: China, with lower regulatory hurdles, is racing to a lead in gene therapy. *Business Week* 2006 March 6.
25. Jia H, Kling J. China offers alternative gateway for experimental drugs. *Nat Biotechnol* 2006; **24**(2): 117–18.
26. Callahan D. *What Price Better Health: Hazards of the Research Imperative.* Berkeley, University of California Press, 2003.
27. Evans JH. *Playing God? Human Genetic Engineering and the Rationalization of Public Bioethical Debate.* Chicago, University of Chicago Press, 2002.
28. Lederer SE. *Subjected to Science: Human Experimentation in America before the Second World War.* Baltimore, Johns Hopkins University Press, 1995; 106.
29. The National Commission for the Protection of Human Subjects of Biomedical and Behavioural Research. *The Belmont Report: Ethical Principles and Guidelines for*

the Protection of Human Subjects of Research. US Department of Health, Education, and Welfare, 1979.

30. See, for example, Angell M. The ethics of clinical research in the Third World. *N Engl J Med* 1997; **337**(12): 847–9.
31. See, for example, Varmus H, Satcher D. Ethical complexities of conducting research in developing countries. *N Engl J Med* 1997; **337**(14): 1003–5.
32. CIOMS. *International Ethical Guidelines for Biomedical Research Involving Human Subjects; Guideline 10.* Geneva, Council for International Organizations of Medical Science (CIOMS), 2002.
33. The Nuffield Council Council on Bioethics. *The Ethics of Research Related to Healthcare in Developing Countries.* London, Nuffield Council Council on Bioethics, 2002.
34. The National Bioethics Advisory Commission. *Ethical and Policy Issues in International Research: Clinical Trials in Developing Countries.* NBAC, Bethesda, MD, 2001.
35. European Group on Ethics in Science and New Technologies to the European Commission. *Ethical Aspects of Clinical Research in Developing Countries*, No. 17. 2003 February 4.
36. Department of Health of South Africa. *Guidelines for Good Practice in the Conduct of Clinical Trials in Human Participants in South Africa.* Department of Health of South Africa, 2000.
37. Indian Council of Medical Research. *Ethical Guidelines for Biomedical Research on Human Participants.* New Delhi, Royal Offset Printers, 2006.
38. Acharya GP, Gyawali K, Adhikari RK, Thaler JL, eds. *National Ethical Guidelines for Health Research in Nepal*, 2001. available at: www.nhrc.org.np/guidelines/nhrc_ethicalguidelines_2001.pdf [last accessed June 4, 2009].
39. Participants in the 2001 Conference on Ethical Aspects of Research in Developing Countries. Moral standards for research in developing countries: from "reasonable availability" to "fair benefits". *Hastings Cent Rep* 2004; **34**(3): 17–27.
40. Participants in the 2001 Conference on Ethical Aspects of Research in Developing Countries. Moral standards for research in developing countries: from "reasonable availability" to "fair benefits". *Hastings Cent Rep* 2004; **34**(3): 17–27.
41. Participants in the 2001 Conference on Ethical Aspects of Research in Developing Countries. Moral standards for research in developing countries: from "reasonable availability" to "fair benefits". *Hastings Cent Rep* 2004; **34**(3): 17–27; at 24.
42. Kelly E, Russell SJ. History of oncolytic viruses: genesis to genetic engineering. *Mol Ther* 2007; **15**(4): 651–9.
43. London AJ, Kimmelman J. Justice in translation: from bench to bedside in the developing world. *Lancet* 2008; **372**(9632): 82–5.
44. Jha P, Mills A, Hanson K, Kumaranayake L, Conteh L, Kurowski C, *et al.* Improving the health of the global poor. *Science* 2002; **295**(5562): 2036–9.
45. London AJ. Justice and the human development approach to international research. *Hastings Cent Rep* 2005; **35**(1): 24–37.
46. National Bioethics Advisory Committee. Volume I: *Report and Recommendations of the NBAC.* In: Ethical and Policy Issues in International Research: Clinical Trials in Developing Countries, 2001; 63.
47. Ponder KP, Srivastava A. Walk a mile in the moccasins of people with haemophilia. *Haemophilia* 2008; **14**(3): 618–20.
48. Pearson S, Jia H, Kandachi K. China approves first gene therapy. *Nat Biotechnol* 2004; **22**(1): 3–4.
49. Corona Gutierrez CM, Tinoco A, Navarro T, Contreras ML, Cortes RR, Calzado P, *et al.* Therapeutic vaccination with MVA E2 can eliminate precancerous lesions (CIN 1, CIN 2, and CIN 3) associated with infection by oncogenic human papillomavirus. *Hum Gene Ther* 2004; **15**(5): 421–31.
50. Corona Gutierrez CM, Tinoco A, Lopez Contreras M, Navarro T, Calzado P, Vargas L, *et al.* Clinical protocol. A phase II study: efficacy of the gene therapy of the MVA E2 recombinant virus in the treatment of precancerous lesions (NIC I and NIC II) associated with infection of oncogenic human papillomavirus. *Hum Gene Ther* 2002; **13**(9): 1127–40.
51. Voltarelli JC, Couri CE, Stracieri AB, Oliveira MC, Moraes DA, Pieroni F, *et al.*

Autologous nonmyeloablative hematopoietic stem cell transplantation in newly diagnosed type 1 diabetes mellitus. *JAMA* 2007; **297**(14): 1568–76.

52. Munro A. Rexin-G(TM), the world's first tumor-targeted gene therapy vector, stymies metastatic cancer. *Medical News Today* 2006 Nov 7.
53. Abdullah J, Isa MN. Analysis of brain tumours suitable for curability via gene therapy in North East Malaysia. *Stereotact Funct Neurosurg* 1999; **73**(1–4): 19–22.
54. Aiuti A, Ficara F, Cattaneo F, Bordignon C, Roncarolo MG. Gene therapy for adenosine deaminase deficiency. *Curr Opin Allergy Clin Immunol* 2003; **3**(6): 461–6.
55. Connock M, Burls A, Frew E, Fry-Smith A, Juarez-Garcia A, McCabe C, *et al*. The clinical effectiveness and cost-effectiveness of enzyme replacement therapy for Gaucher's disease: a systematic review. *Health Technol Assess* 2006; **10**(24): iii–iv, ix–136.
56. National Hemophilia Foundation. *Financial and Insurance Issues*. www.hemophilia.org/NHFWeb/MainPgs/MainNHF.aspx?menuid=34&contentid=24&rptname=bleeding; 2006.
57. Lu ZJ, Comanor WS. Strategic pricing of new pharmaceuticals. *Rev Econ Stat* 1998; **80**(1): 108–18.
58. Harper M. Avigen leads gene therapy charge. *Forbes.com* 2000 (November 28).
59. Fontes EMA, Amorim L, Carvalho SM, Farah MB. Hemophilia care in the state of Rio de Janeiro, Brazil. *Rev Panam Salud Publica* 2003; **13**(2–3): 124–8.
60. Jia H. Controversial Chinese gene-therapy drug entering unfamiliar territory. *Nat Rev Drug Discov* 2006; **5**(4): 269–70.
61. China's war on cancer. *Red Herring: The Business of Technology* 2006.
62. Hogerzeil HV. Rare disease and essential medicines: a global perspective. *Int J Pharm Med* 2005; **19**(5–6): 285–8.
63. Reidenberg MM. Are drugs for rare diseases "essential"? *Bull World Health Organ* 2006; **84**(9): 686.
64. World Health Organization. *The Selection and Use of Essential Medicines*. WHO Technical Report Series 933, 2005.
65. World Health Organization. *Expert Committee Report*, 2005 June 27.
66. Stonebraker JS, Amand RE, Nagle AJ. A country-by-country comparison of FVIII concentrate consumption and economic capacity for the global haemophilia community. *Haemophilia* 2003; **9**(3): 245–50.
67. Ford N, Darder M. Registration problems for antiretrovirals in Africa. *Lancet* 2006; **367**(9513): 794–5.
68. Lanjouw JO. *Patents, Price Controls, and Access to New Drugs: How Policy Affects Global Market Entry*. National Bureau of Economic Research, NBER Working Paper No. 11321, 2005.
69. Clemens J, Jodar L. Introducing new vaccines into developing countries: obstacles, opportunities and complexities. *Nat Med* 2005; **11**(4 Suppl.): S12–15.
70. Lassarat S, Jootar S. Ongoing challenges of a global international patient assistance program. *Ann Oncol* 2006; **17**(Suppl. 8): viii43–viii46.
71. Strom S, Fleischer-Black M. Drug maker's vow to donate cancer medicine falls short. *The New York Times* 2003 June 5.
72. Calabro S. Genzyme: The price of success. *PharmExec.com* 2006.
73. Mackie JE, Taylor AD, Daar AS, Singer PA. Corporate social responsibility strategies aimed at the developing world: Perspectives from bioscience companies in the industrialized world. *Int J Biotechnology* 2006; **8**(1/2): 103–17.
74. Anonymous. When a drug costs $300,000. *The New York Times* 2008 March 23.
75. Pollack A. Drug maker stays close to doctors and patients. *The New York Times* 2008 March 16.
76. Pollack A. Cutting dosage of costly drug spurs a debate. *The New York Times* 2008 March 16.
77. Anand G. Why Genzyme can charge so much for Cerezyme. *The Wall Street Journal* 2005 November 16.
78. Scherer FM, Watal J. Post-trips options for access to patented medicines in developing nations. *J Int Econom Law* 2002; **5**(4): 913–39.
79. Barton JH, Emanuel EJ. The patents-based pharmaceutical development process. *JAMA* 2005; **294**: 2075–82.

80. Plahte J. Tiered pricing of vaccines: a win-win-win situation, not a subsidy. *Lancet Infect Dis* 2005; **5**(1): 58–63.

81. Batson A. The problems and promise of vaccine markets in developing countries. *Health Aff* 2005; **24**(3): 690–3.

82. Grace C. *Equitable Pricing of Newer Essential Medicines for Developing Countries: Evidence of the Potential of Different Mechanisms.* London, London Business School, 2003; 1–70.

83. Kapczynski A, Chaifetz S, Katz Z, Benkler Y. Addressing global health inequities: an open licensing approach for University innovations. *Berkeley Technology Law J* 2005; **20**(2): 1031–114.

84. Berndt ER, Hurvitz JA. Vaccine advance-purchase agreements for low-income countries: practical issues. *Health Aff* 2005; **24**(3): 653–65.

85. Chase M. Malaria trial could set a model for financing of costly vaccines. *The Wall Street Journal* 2005 April 26.

86. Flory JH, Kitcher P. Global health and the scientific research agenda. *Philos Public Aff* 2004; **32**(1): 36–65.

87. Mathers CD, Loncar D. Projections of global mortality and burden of disease from 2002 to 2030. *PLoS Med* 2006; **3**(11): e442.

Chapter 9

Great Expectations and Hard Times: expectation management in gene transfer

Introduction

In 2001, a report in *Nature Genetics*[1] raised the possibility that a rare, hereditary form of blindness, Leber's congenital amaurosis (LCA), might soon have a cure. A team of researchers led by Jean Bennett of the University of Pennsylvania's Scheie Eye Institute had successfully restored vision to three dogs with LCA. According to news stories, the report "electrified" families with the disease; said one mother of an LCA child, "we are bursting at the seams."[2] Word spread like "wildfire," according to Bennett, who received hundreds of inquiries from expectant parents.[3]

The next year, the Alliance for Eye and Vision Research and the Foundation Fighting Blindness dispatched a team of researchers to meet with lawmakers during hearings on NIH appropriations. Traveling with them was Lancelot, one of the Briard mix dogs whose blindness had been partially corrected. One member of the research team, Cornell's Gustavo Aguirre, directed lawmakers to note the dog's posture: Lancelot always stood to one side – he favored his right eye because his left had served as the uncorrected experimental control.[4] "If any of the investigators in the study had not received funding from the National Eye Institute/NIH," Aguirre stated, "this amazing breakthrough would never have come to pass. Without increased funding, much of this promise will languish on the laboratory bench."[5]

The hearings followed one year on the heels of the first leukemia diagnosis in the Paris X-SCID study, and two years after Jesse Gelsinger's death. In addition to embodying the promise of vision research, Lancelot's eye also projected a favorable image for the embattled field. "This is the promise of gene therapy," Aguirre told a reporter in 2005. "Making the lame walk and the blind see."[4] Aguirre was, of course, speaking literally. But his words carried figurative meaning as well: through Lancelot, skeptics might come to share his vision for gene therapy.

Seeing a seeing dog is believing, and Lancelot soon became one of the most circulated witnesses for gene transfer. He sat at the feet of Congressman C. W. Bill Young, who chairs the House Appropriations Committee.[6] He also laid down for then Secretary of Health and Human Services Tommy Tompson.[7] His photograph adorns the offices of the Foundation for Fighting Blindness, which co-sponsored the study.[8] Following the third leukemia in the Paris X-SCID trial, former ASGT president Savio Woo directed reporters to Bennett's work: "she has a bunch of dogs walking around with their heads tilted."[9] And in a 2007 meeting of the ASGT, the organization "highlight[ed] some of the latest and most promising developments in gene- therapy products"[10] by inviting Bennett to present reporters with a video of Lancelot navigating a dim room.

Yet Bennett's hopeful message at the ASGT meeting sounded a cautionary note. Bennett described having received a letter from an LCA parent. It read: "I recently heard about your successful studies involving treatment of blindness and I wonder whether you can please help my 16 year old girl? I have brought her to every specialist I can and am not satisfied with the results. Would it be possible for me to bring her to your lab this week to receive the treatment? I will pay whatever it takes to make my Judy happy. Please call me."

Translational research and performance

Where some might see in the profuse display of Lancelot a process of raising public awareness about important research, others might regard it as just the kind of stagecraft that leads patients and policy-makers toward unsustainable expectations. Regardless of where one stands, translational research involves public theater – trials are, in the words of one gene-transfer researcher, "highly visible because of public, commercial, and political interest"[11] – and scientists actively manage expectations about their research.

Investigators and academic medical centers must make their work visible to draw support from biotechnology firms, public funding agencies, regulatory bodies, and prospective research subjects. Yet they must do so carefully, lest expectations expose the limits of their powers. Moreover, researchers can scarcely hide: media interest in cutting-edge research is intense, and patients – along with their families and advocates – often track developments closely. Even if they would prefer otherwise, translational research involves public engagement.

In this chapter, I present three arguments. First, researchers manage expectations of various publics. Researchers are not the only parties that manage expectations around translational research. Members of the media, biotechnology firms, disease advocates, and activists also play roles in shaping expectations around translational research.* My focus in this chapter, however, will be the role of investigators – those with the greatest informational resources and who are bound by certain professional commitments.

Second, I argue that scientists manage expectations through a series of activities that involve theatrical and nonverbal elements. Sociologist Erving Goffman famously argued that people use dramaturgical techniques, like controlling information flow, coordinating staging with fellow performers, policing the "curtain" between front and backstage, segregating audiences, and realigning after gaffes, intended to control how others perceive and define a situation.[16,17]† The display of Lancelot exemplifies how this might apply to scientists. The nomenclature provides an example: the original *Nature Genetics* publication never referred to Lancelot by name, but instead by laboratory labels like BR29, BR33, BR47. Before Diane Sawyer on *Good Morning America*, however, the dog became "Lancelot," presented as if he were a faithful pet rather than a reliable animal model. Why not show the other two dogs? Lancelot's sister had a "testy temperament;" the other was euthanized to perform further experiments.[7]

The third thesis presented here is that, though expectation management and its staging‡ are not in themselves ethically problematic, they *are* ethically consequential. Expectation

* For policy and ethical analysis of how the media convey scientific discovery to publics, see Nelkin D., *Selling Science*,[12] Friedman *et al.*, *Communicating Uncertainty*.[13] For patient advocates, see *When Science Offers Salvation*, Chapter 7.[14] The business ethics literature on biotechnology communications tends to be less developed. Nevertheless, several short articles that are focused on public relations offer insights about expectation management. One example is Phillips & Beckman.[15]

† Goffman's work has been productively applied to science and technology studies by Steven Hilgartner in his analysis of deliberations over diet and health at the National Academy of Sciences. See Hilgartner S., *Science on Stage*.[17] Formulating expectation management as a type of public theater is somewhat analogous to how many commentators describe the process of risk communication. For one example of this in the realm of biotechnology, see Brauerhoch *et al.*[18]

‡ Goffman is agnostic about the moral implications of presentation of self. Ethicists don't have that luxury. For now, however, I wish to emphasize that impression management, related to deception, is not

management isn't ethically problematic because it is one of the primary vehicles by which investigators vouch for the promise of their research. It is ethically consequential, however, for several reasons. Trial volunteers often arrive at consent discussions already having decided to enroll.[19] Consent decisions are therefore shaped by the expectations volunteers bring with them. Moreover, expectations have major consequences in determining whether and to what extent scientific research programs are funded, authorized, and supported.

Expectation management can become problematic where it embeds messages or claims that lead to public expectations that are grossly inconsistent with consensus expert opinion. It is also ethically problematic where it stymies processes by which members of the public and researchers can form mutual understanding. Determining when either occurs is no small task, and the current chapter is intended as an exploration of the nature and consequences of two prevailing modes of expectation management in translational research.

The first, "Great expectations," describes a series of activities in which researchers project the promise of their research. The most objectionable form of this mode is often referred to as "hype," "sensationalism," or "hyperbole," though for reasons outlined later in this chapter, I prefer to avoid these labels. The second mode of expectation management, "Hard times," presents adversity when research fields stumble, or where expectations go unredeemed.[§] The chapter concludes with a series of modest ethical and policy prescriptions.

Great expectations

Translational researchers must create the future out of nothing: decisions to pursue the development of novel interventions are propelled by beliefs about promise rather than current realities.[¶] Expectant projections from preclinical studies help others to imagine the implications of esoteric scientific findings; a society unable to imagine future applications would not likely make serious investments in biomedical research.

Building expectations is thus part of a critical process by which researchers enlist support from sponsors, policy-makers, patient communities and the public.[21] Generating expectations is especially important for helping policy-makers and members of the public form reasoned opinions and policy in contested research areas like human genetic engineering (in the early 1980s) and embryonic stem cell research (today).

Nevertheless, the process of building expectations is fraught with peril, and gene transfer has frequently been excoriated for "giv[ing] a continuous positive 'spin' that is unusual for most medical research" (see Box 9.1).[22]

---

necessarily ethically suspect. That the host tidies his living room to leave a favorable impression with dinner guests does not make him dishonest.

[§] The division is not a neat one: it hardly exhausts the types of expectation researchers manage. Moreover, many episodes contain elements of both. The presentation of Lancelot, for example, at once generated great expectations by showing the promise of eye research, while dispelling hard times by drawing attention away from the field's troubled past.

[¶] Perhaps a more rigorous way of stating this is that, at the point where a novel translational research project is initiated, the future is more speculative than where a currently existing intervention or class of interventions is modified in an incremental manner. In the latter instance, members of the public and research communities are familiar with the properties and capacities of the technology. In the translational context, the future is less easily projected from current experience. For a further elaboration, see Brown & Michael[20] and Hedgecoe.[21]

Box 9.1 Hard sell: are gene-transfer researchers entirely to blame for expectation?

Condemnation of gene transfer has become so ritualized within science, policy, and ethics circles that it warrants scrutiny. Those who condemn gene transfer should avoid the temptation of viewing the field as a monolithic entity. Numerous senior figures in the field have long been critical of hasty trials (e.g. "the idea of doing human gene therapy is now more acceptable. I don't know why. It is not the success of the experiments" (1992)[23]) or warned of "overemphasizing... potential benefits to an unrealistic level and... minimizing or disregarding their potential problems"[24]

Even some of the field's most ambitious entrepreneurs have at times reflected about such issues. One example is a publication on controlling therapeutic misconception; the paper's senior author was Ronald Crystal, an ambitious researcher who has occasionally been the target of criticism.[25]

Perhaps most importantly, expectation is less a "pipeline" than a "network" phenomenon. "Overselling" is instigated, abetted, amplified, and propagated at numerous social nodes (e.g. venture capitalists, academic medical centers, disease advocates, journalists, research institute directors, press officers, medical journal editors).

I sometimes wonder whether the condemnation of gene transfer is more a reflection of the field's cohesiveness as a category – and its relationship with another allegedly "oversold" field, genetics – than any unusual overzealousness. That is, there are numerous isolated examples of investigators overzealously presenting their research findings, or orchestrating elaborate campaigns to garner public and private support. One example is the selling of interferon therapies in the 1970s (see especially chapters 5–7 of Pieters's *Interferon*[26]), another is artificial retinas.[27] Gene-transfer strategies are multiple, but because they are related to each other by a categorical term (i.e. "gene transfer") and a set of techniques that are conspicously distinctive, it is much easier to perceive a pattern, and perceptions of one investigator's conduct tend to reinforce similar perceptions of other investigators performing research within the same category.

Three episodes from the early days of gene transfer show certain signature methods by which researchers and others generate expectations that can founder on complexity and social dynamics. The first involves the machinations of W. French Anderson in the lead-up to the first ever gene-transfer trial. According to his own account, Anderson had become convinced that gene transfer was ready for clinical trials since the early 1980s. But a scientific review committee within the RAC, the Human Gene Therapy Subcommittee (HGTS), was skeptical, and when Anderson proposed human studies it twice turned him down on account of insufficient supporting data. Anderson attributed the rejections to a "'no gene therapy' mindset." Undaunted, he and colleagues Steven Rosenberg and Michael Blaese devised a two-pronged strategy to overcome opponents on the HGTS. First, they would obtain further supporting evidence, but withhold it from the committee until weeks before its meeting, using the pretext of protecting publication rights. This would prevent the body's skeptics from having sufficient time to pick over its deficiencies. Second, they would present a "carefully planned emotional appeal." The team reasoned that most opposition on the committee originated from scientists like Richard Mulligan and Scott McIvor. Instead of addressing their concerns, the team would appeal to others on the committee. Rosenberg was scripted to say "485,000 Americans died of cancer last year – that means that one person dies of cancer every minute." But when he missed his cue, Anderson waited for a strategic opening. "I stood up and said (what I had been rehearsing for a week): 'Perhaps the RAC members would like to visit Dr. Rosenberg's cancer service and ask a patient who has only a few weeks to live: "What's the rush [to begin the first human gene transfer experiment]?" A patient dies of cancer every minute in this country. Since we began this discussion 146 minutes ago, 146 patients have died of cancer.'"[28] The tactics worked, as three members – none of whom were molecular biologists – voiced tentative support for the proposal. With little warning, the RAC's chair moved for a vote, and the protocol was approved over the objections of several scientists. Days later, the Anderson team celebrated with a banquet, at which laboratory members staged another performance – skits mocking various panelists on the RAC.[28]

James Wilson proposed the second ever gene-transfer trial involving a genetic illness, familial hypercholesterolemia (FH). His protocol involved a severe procedure in which as much as 15 percent of the FH volunteer's liver would be removed, genetically modified, and returned to the volunteer. First and foremost in Wilson's presentation of the procedure to the RAC was his conjuring of the off-stage patient,[29] in this instance, Stormie Jones, an energetic Texas girl with FH who had famously received the first ever heart-liver transplant.[30,31] The juxtaposition not only evoked the gravity of the disorder, but also highlighted the protocol's affiliation with heroic organ transplantation.[††] It also provided a convenient "hook" for journalists covering the proceedings.[‡‡] Unlike Anderson's study, Wilson had compelling evidence of efficacy in an animal model, and the RAC needed little convincing; indeed, upon approval, RAC chair Gerard McGarrity stated (according to meeting minutes) "the action signifies a new attack on yet another genetic disease, and represents an important advance in human gene therapy of genetic diseases."[29]

The trial approved, the University of Pennsylvania (which inherited it from the University of Michigan after Wilson moved there in 1993) characterized the procedure as "gene therapy treatments" and "treated" the first "patient" on June 5, 1992.[36] Using unusually florid language for a scientific publication, the study's authors proclaimed it "the first report of human gene therapy in which stable correction of a therapeutic endpoint has been achieved."[37§§] The team held a press conference at which they exhibited their study's first volunteer. The *New York Times*'s Natalie Angier reported "her blond hair swept back and her prim white blouse buttoned up to the collar, she said she had felt 'very well' since the operation in 1992. Speaking through an interpreter in her native French, she said: 'I feel very well physically and morally. I feel I can do more physical activity, like skiing, dancing and other social activities.'"[39] The *Washington Post* described the 30-year old as standing "flanked by her doctors, including James M. Wilson," and "looking trim and healthier than many of the reporters who attended."[40] And the associated press quoted her "I am certainly going to live 90 years of age, and more, possibly."[41] The papers described the study as "show[ing] all the signs of a real, if modest,

[††] In light of the widespread representation of the trial as a therapeutic experiment, it may be worth addressing the question of whether investigators were unethical in portraying the experiment as therapeutic. Certainly, the study design does not appear to have involved procedures, like subtherapeutic dosing, that would frustrate therapeutic objectives. Yet, the fact that the NIH – a USA-based research agency – picked up the $100 000 tab for this highly invasive procedure ought to tell us something about the primary motivations of sponsors and the research team.[32]

[‡‡] The link with Stormie Jones was picked up in several news reports of the trial's approval at the RAC.[33–35]

[§§] Note that in letters and an accompanying "News and Views" piece, various researchers challenged the claim, describing the improvements as "modest," raising the possibility that effects were attributable to either a change in lifestyle for the volunteer or a placebo effect, and pointing out the Wilson team's failure to demonstrate that clinical improvement was attributable to synthesis of the new gene rather than stimulation of the volunteer's partially functional one.[38]

triumph" and a "landmark," respectively.¶¶ Other geneticists and gene-transfer researchers hailed the result as "work[ing],"[43] "a technical tour de force,"[44] "proof that this approach, which has been talked about so much and has stirred a bit of controversy, can do what it's supposed to do,"[45] and "further strengthen[ing] the whole premise that gene therapy is going to revolutionize science and medicine."[46]

The third gene-transfer trial involving a genetic disorder was NIH researcher Ronald Crystal's. The cystic fibrosis protocol was one of three presented to the RAC when it met in December 1992, but Crystal was the first to initiate his study and publish results. A year earlier, Crystal and collaborators had published a study showing that they could use adenoviral vectors to transfer the CFTR gene (mutations of which causes cystic fibrosis in human beings) to the lungs of cotton rats and observe gene expression. Describing the study in the *Washington Post*, Crystal stated "I have no doubt at all that if we were to put this in the lungs of patients, we would correct the cystic fibrosis now ... The major question is whether it's safe." He added "I really don't foresee any problems ... We have some hurdles down the road, but it's remarkable how fast this field is moving. I think we're going to cure this disease."[47] Disease foundations – which had largely stood on the sidelines in previous studies – embraced Crystal's enthusiasm. Responding to another study led by Crystal a year earlier, the executive vice-president for medical affairs at the Cystic Fibrosis Foundation said "we can now look into the eyes of young people suffering from this disease and promise them a life that is going to be very different from the way it has been in the past."[48] Like Crystal,*** this disease advocate saw gene transfer as a potential cure: "We're very excited – [gene transfer is] an opportunity to provide a truly genetic cure of a genetic disease."[52]

The mere approval of Crystal's protocol by the RAC was described by reporters as "signal[ing] a new era for the revolutionary technique."[53] Executive secretary of the RAC Nelson Wivel (himself a gene-transfer researcher) characterized the approval as "... a major, watershed event."[54] When the trial was initiated, journalists described it as "fulfilling hopes that had gathered locomotive force in the past several months."[55] Notwithstanding a series of cautionary statements offered by Crystal, his own expectations were clearly high going into the study: he reportedly saved the syringe used to administer vector to the lungs of the first volunteer as a memento.[55] When a competing team announced the results of their studies, the *Boston Globe* reported "the advance ... represents the first time scientists have reversed a genetic disease by introducing the missing gene into a living patient."[56] Responding to a competing study that corrected the defect in nasal passages, National Heart, Lung, and Blood Institute (NHLBI) director Claude Lenfant stated "I would be very disappointed if in a decade or a little bit more we have not solved all the problems [of gene therapy for cystic fibrosis]".[56]

¶¶ Once the trial had been completed in five volunteers, the procedure's efficacy was less apparent. R. Sander Williams, in a commentary accompanying the full publication, likened it to recently published disappointing results in cystic fibrosis and muscular dystrophy trials. "The results are disappointing: three patients manifested measurable changes in LDL catabolism, but plasma LDL cholesterol levels remained at severely atherogenic levels ... the efficacy of drug therapy was unchanged after gene transfer in these patients."[42]

*** Other examples: "Experiments carried out in our laboratory demonstrated that this was a feasible approach, with minimal risk to potential subjects. We quickly recognized that we had developed a strategy with the potential to cure this devastating disorder, and we moved rapidly to evaluate this strategy in individuals with cystic fibrosis.;"[49] "The time is right to cure this disease, and while I don't think we're going to be able to do that immediately, we now have the potential for a cure"[50]. The sentiments were

Appraising great expectations

Do the above performances, and others like them, deserve condemnation? If so, where do they cross from enthusiasm to deception, from respect for their audiences to contempt, or from what one philosopher called communicative action to strategic action?

The standard view would tend to conceptualize expectation as a problem of information and, in particular, truthful disclosure. Thus, various commentators suggest that "scientists who provide the information [that leads to "sensationalized hope"] must share the blame,"[57] and urge researchers to "restrain the promotional tendencies that lead to ... oversell,"[58] or to "communicate research results accurately."[59] Such positions are embodied in several formal ethics policies as well. For example, the American College of Physicians states that researchers should "use ... precise and measured language" and "avoid raising false public expectations or providing misleading expectations."[60] The Medical Research Council of India states "Researchers have a responsibility to make sure that the public is accurately informed about results without raising false hopes or expectations."[61]

While these positions are all perfectly reasonable, they also have certain limitations in how they might diagnose the cause and consequences of expectations in the three episodes above. Consider the view that expectation is a problem of misinformation. Certainly, there are numerous anecdotal reports of biotechnology firms – and persons in their pay – deliberately distorting experimental results in order to attract greater investment. For example, James Wilson stated that leaders of a gene-therapy company "try to put as positive a spin as you possibly can" on every step of the research process ... "because you have to create promise out of what you have – that's your value."[62] Few would defend such conduct.[†††] Nevertheless, in only one of the three episodes above did researchers clearly use their performance in an instrumental way. Anderson, Rosenberg, Wilson, and Crystal follow in the tradition of clinical champions (see Chapter 3). Like all clinical champions, they probably believed in their cause; their optimism and conviction are what drives them to embark on risky and uncertain projects. Renée Fox and Judith Swazey describe a similar type of optimism.[63‡‡‡] Moreover, though the above episodes contain numerous examples of hyperbole, many statements were also tempered. For example, at the outset of cystic fibrosis trials, Crystal "urged families of CF patients not to raise their expectations too high. 'There is a small chance that this will significantly benefit the patients who will be treated.... However, we're only giving the treatment once. We don't know how long it will last.'"[53§§§] Likewise, Wilson often qualified his claims about the FH results, stating "we have not yet cured familial hypercholesterolemia," or referring to the results as a "partial correction."[39]

echoed by NHLBI head Claude Lenfant: "If this approach works, we may finally see a cure for this lethal disease."[51]

††† The neoliberal critique would posit that such practices lead to market inefficiencies because of information asymmetries. It is for this reason that agencies like the US Securities and Exchange Commission and Food and Drug Administration regulate commercial speech.

‡‡‡ One recurrent theme in an emerging "sociology of expectation" literature is the concept, first articulated by Donald MacKenzie, of the "uncertainty trough:" producers of knowledge and end users acknowledge uncertainties surrounding a new technology, whereas intermediaries that neither produce nor use the technology are less aware of uncertainties. I find it difficult to reconcile the concept with innovative clinical research and practice, where producers of knowledge tend to express the greatest level of certainty.

§§§ Other examples of moderating quotes from Ronald Crystal include: "It's a demonstration that you can correct the abnormality in the nose, but the disease is not in the nose."[64]; "To think we're going to cure cystic fibrosis in a year is naïve. I'm not discouraged, but this is going to take time and people shouldn't

One of the most infamous recent examples of medical sensationalism occurred in 1998, when the Sunday *New York Times* ran a front-page story describing Judah Folkman's discovery of a class of cancer drugs called angiogenesis inhibitors.[67] The report quoted Nobelist James Watson saying "Judah is going to cure cancer in two years," and then National Cancer Institute director Richard Klausner as stating that he was "putting nothing on higher priority than getting this into clinical trials". The next morning, shares of Entremed, a biotechnology firm founded by Folkman, increased six fold,[68] and when Harvard initiated trials on one of the products – endostatin – over 1400 cancer patients sought entry for three slots; one wealthy study candidate even offered to buy a controlling stake in Entremed.[69] Was this a case of researchers and journalists "gone wild"? Perhaps, although the original news item also described Folkman as "cautious about the drugs' promise. Until patients take them, he said, it's dangerous to make predictions. All he knows … is that 'if you have cancer and you are a mouse, we can take good care of you.'" Said Folkman elsewhere in the story: "going from mice to people is a big jump, with lots of failures." Clearly there is much more behind expectation than intentional misinformation of the public.

Defining expectation as a problem of information and accuracy also has epistemological problems. There exist no objective standards by which to demarcate inflated and realistic scientific claims. As such, many prescriptions offer little guidance. For example, the Association of Health Care Journalists (AHCJ) "Statement of Principles," urges reporters to "distinguish between advocacy and reporting." and avoid "report[ing] bold or conclusive statements about efficacy in Phase 1 trials."[70] As indicated above, all scientists are, in some sense, advocates of their research programs, paradigms, and interpretations of data; advocacy and reporting often blend seamlessly. The suggestion that "journalists, scientists, and physicians … have professional responsibilities to meet standards of truthful disclosure"[59] has similar limitations. The policy can be interpreted in three ways. First, scientists might avoid deliberately deceiving the public. Scientists do occasionally deceive their publics – often with good intentions. Ron Worton, who first isolated genetic sequences involved in Duchenne's muscular dystrophy, describes projecting a cure for the disease five years from the discovery. Thirty years later, he admitted regretfully "I knew in my heart I wasn't being completely truthful."[71] Nevertheless, in most cases researchers who need to be told not to lie are probably the least likely to respond to such exhortations. A second way of interpreting "truthful disclosure" policies like the AHCJ's might be that researchers should employ a certain affect management in which they "curb their enthusiasm" when communicating a research findings to the public by using subdued language and emphasizing limitations.[72] This form of truth telling is potentially self-defeating, since it would seem to ask the investigator to describe circumstances in ways other than those in which he or she sees them. A third interpretation of the truth-telling prescription might encourage the investigator to describe a study as he or she sees it, but asks him or her instead to believe in a more subdued reality. But clinical champions are constitutively optimistic;[§§§] expectation provides an important vehicle through which clinical champions actualize ambitious research objectives.

---

have unrealistic expectations"[65]; "It's hard to reproduce what evolution has already figured out how to do so well."[66]

[§§§] Renée Fox and Judith Swazey describe this phenomenon in their account of transplant surgeons: "This pioneer syndrome is closely related to an emphasis on optimisim in the face of uncertainties, limitations, and high mortality rates. This 'accentuate the positive' tendency is more pronounced in some of these surgeons … they contend that 'seeing the glass as half full rather than half-empty' is both harder and more necessary for them, given the stress and complexity that operating on the heart

Instead, nonverbal, symbolic, and contextual aspects of the above episodes would seem to matter as much if not more in how expectations are generated and transmitted. Sociologist Dorothy Nelkin, quoting the political scientist Harold Laski, stated that "the real power of the press . . . comes from its ability to surround facts by an environment of suggestion which, often half consciously, seeks its way into the minds of the reader and forms his premises for him."[75] The same might be said about the researcher's public performance; a more incisive analysis and ethics of expectation should attend both to actual claims as well as the "environment of suggestion." In the episodes described above and many others subsequent, several features stand out.

Therapeutic drama: Translational studies are presented as a therapeutic drama. Anderson and Rosenberg were brazen in mobilizing their patients' needs to generate expectations,[76] as their procedure had no therapeutic impetus at all (it was aimed at following the migration of lymphocytes to tumors). Anderson established the exigency of his study by capitalizing on a longstanding cultural ambivalence about the aims and value of clinical experimentation.[77] Certain committee members viewed his protocol as an experiment, and thus judged it on the merits of evidentiary justification, design, and methodology. Rather than trying to appeal to them, Anderson and his coworkers presented themselves as a tight-knit team of dedicated physicians caught up in the all too familiar predicament of clinical uncertainty.**** Wilson also sought to link his procedure with drastic therapeutic interventions, borrowing the stagecraft of transplant teams.[78]

This therapeutic drama often takes subtle forms in which what is gestured at betrays what is stated. In 1999, researchers at Ohio State University initiated the first ever gene-transfer study involving muscular dystrophy. The protocol involved injecting a foot muscle with vector while the same muscle in the opposing foot received a sham injection. Because the study did not involve systemic administration, it had no therapeutic value whatsoever. In a press release issued by Ohio State, the study's lead investigator made this clear: ". . . we want everyone to know it's only a first step. We don't expect [the volunteer] or any other patient to directly benefit from this trial." Yet the release also contained the statement from the study's first volunteer, Donavon Decker, "I went from not knowing if there would ever be a cure to where we think now that it's just right around the corner . . . it's kind of been a long waiting period, but this is the first step of the cure."[79] The procedure was allegedly timed to coincide with the annual telethon for the Muscular Dystrophy Association, and was videotaped for posting on the organization's web page.[80] Here, as in Crystal's first cystic fibrosis study, the act of performing a study that involves sick patients became a display of compassion, decisive action, and therapeutic engagement – qualities we normally associate with medical care – even where the individual protocol had no therapeutic value.

---

entails."[73] None other than Steven Rosenberg once stated "Cancer patients deserve optimistic doctors. And I'm optimistic."[74]

**** In 1995, the experimentalists would reassert their views, when NIH head Harold Varmus assigned two distinguished researchers, Stuart Orkin and Arno Motulsky, to appraise the field Orkin–Motulsky report. It should be noted that mere professional affiliation does not determine which way an individual will view. Still, it is striking that many of the main protagonists pushing gene transfer into trials have been MD, while the most vocal critics from the inside (Richard Mulligan and Inder Verma) hold PhDs. Varmus – an MD – reportedly selected Orkin and Motulsky for their their "stature in the scientific community." It is striking, then, that both Orkin and Motulsky are MDs; it is possible that their selection was intended to reinforce the credibility of the report within medical circles.

Reduction of complexity and contingency:[82] Conditional observations made in controlled experimental contexts are projected to messy, more generalized contexts, and diseases are often reduced to solitary phenomena. For instance, Crystal and various other commentators often seemed to reduce cystic fibrosis, which is a multi-system disease, to a pulmonary disorder such that gene transfer to the lung came to be seen as a cure instead of what Lewis Thomas might call a "halfway technology." True, correcting the pulmonary defects would dramatically extend survival and quality of life for persons with cystic fibrosis. But it would likely be accompanied by a sharp rise in cystic fibrosis-related diabetes, which is related to pancreatic dysfunction and is itself associated with increased mortality and pulmonary complications.[83] Similarly, though immunological deficiencies are the cause of mortality in ADA-SCID, the disease is also associated with various neurological, skeletal, hepatic, and renal complications. Correcting the immunological deficits represents a tremendous advance, one that easily justifies the expenditure of energy toward this rare disease. But it isn't quite accurate to describe this as a "cure."[84] More recently, researchers and others have frequently described Parkinson's disease gene-transfer interventions as potential cures. As of this writing, all interventions tested have aimed at correcting dysfunctions in the nigrostriatal system; a best-case scenario for these studies would eliminate the most visible and familiar features of the disease, namely tremor and bradykinesia (an inability to initiate movement). But the brain pathology of Parkinson's disease involves multiple systems,[85] and such strategies would probably have little impact on the progression of symptoms like psychosis and dementia.

Environment: The environment in which investigators present their research often drives escalation of expectation. For instance, credible commentators like Claude Lenfant, Francis Collins, and Nelson Wivel, whose relationship with the studies was at least arm's length, often amplified the expectant claims of investigators. Patient advocates also frequently promote such claims, as do academic medical centers. Less noticed is the role of medical journal editors. On the publication of Wilson's trial, *Nature Genetics* published an editorial stating "... the crucial question is whether such a major operation (at an estimated cost of $75 000) has had the desired results. After 18 months of careful monitoring, the answer, reported on page 335 of this issue ... seems to be a cautious 'yes.'" An accompanying commentary ran the title "Heroic Gene Surgery."[38] Elsewhere, journal editors have occasionally asked authors to replace subdued manuscript titles with more exciting ones. Dubowitz describes one instance where a journal editor requested a title change for a preclinical gene-transfer manuscript involving gene transfer for muscular dystrophy.[80] Any prescription on expectation management should reflect how the context of researcher interactions with publics might amplify certain interpretations or dampen others.

Hard times: managing adverse expectations

Whether because of its high visibility, mismanagement, bad luck, or the unpredictability of cutting-edge research, few fields have had to work as hard as gene transfer to manage and overcome adverse publicity and deep skepticism from within the scientific community.

As already recounted, Anderson and colleagues labored to overcome skeptics when they proposed the first two trials. Five years later, the field's reputation suffered a setback when Arno Motulsky and Stuart Orkin presented their critical report to the NIH.[86] But probably the most conspicuous and damaging adversities for the field have been those that unfolded in the public arena.

The most significant was the death of Jesse Gelsinger, which garnered headlines for months and came to symbolize all that ailed gene transfer (see Chapter 3). In the aftermath, many looked to the Paris X-SCID study for vindication of gene transfer.[‡‡‡‡] This then suffered a series of setbacks when volunteers developed lymphoproliferative disorders (see Chapter 4). Many further looked for vindication with the publication of preliminary data from Katherine High's and Mark Kay's trial testing the use of AAV vectors for hemophilia B.[87] Theodore Friedmann's address to the AAAS meeting in February 2000 described this trial as having "the feel of being correct."[90] Excitement about this study, however, was diminished after volunteers in a follow-up study showed increases in serum transanimases (indicators of liver toxicity) and vector sequence contamination of semen.[91] Shortly thereafter, the private sponsor, Avigen, sold its hemophilia gene-transfer program to Genzyme.[92,93]

The headlines in a 2005 *Nature Biotechnology* report captured the way many viewed the field: "Gene Therapy – Cursed or Inching Toward Credibility?" And as private investment waned, the volume of human trials declined. In the face of so many disappointments, gene-transfer researchers developed various strategies and discourses to shore up the confidence of various publics.[94,95] "Hard times" narratives in gene transfer tended to take four forms.

Historical precedent: Many leading spokespersons for gene transfer pointed to other medical innovations whose life-saving potential was only realized after many years of technical adversity. Examples included recombinant proteins,[96] interferon,[97] insulin,[96] monoclonal antibodies,[99–101] bone marrow transplantation,[101] organ transplantation, and the use of alkylating agents in the treatment of cancer.[102] One researcher's thoughtful presentation before the RAC shows just how elaborate the "historical precedent" argument can be. The presentation provided four historic examples, reaching back to 1948, of life-saving medical interventions that showed significant mortality and morbidity on initial attempts.[103] A related trope was the suggestion that such setbacks are inevitable in research: as one researcher put it, "that's why it's called research. We don't know what's going to happen. And in general, it's not a surprise that these kinds of things will happen."[104]

Reinterpreting disruptions: Evidence never speaks for itself, and gene-transfer researchers have sometimes publicly read favorable meanings into otherwise unfavorable experimental outcomes. Thus, while scientists and others outside of gene transfer interpreted the cases of leukemia in the Paris X-SCID study as an indication of the platform's limitations, many inside the field represented these same events as tragic but nevertheless hopeful affirmations of progress. For example, one commentator opined "it's no surprise that the more effective the gene-therapy transfer becomes, the more we're going to see adverse consequences that are intrinsic to that technology."[105] Another stated "for the first eight or nine years of gene therapy, when gene delivery wasn't efficient enough to get an effect, we also didn't see side effects. But every powerful technology has powerful side effects, so naturally there's going to be side effects."[106]

Denial and defensiveness: As one reporter observed, "the field's researchers get a bit defensive when discussing the many problems that have plagued it."[99] One example occurred when Erika Check, a reporter for *Nature*, published a news item on a study documenting a tendency of AAV vectors to insert near actively transcribed genes. The article's strongest assertions were in the headlines ("Harmful potential of viral vectors fuels doubts over gene therapy"), as well as the lead's statement that the study dealt "the troubled field of gene therapy … a fresh blow" and a sentence stating "the finding suggests that the vector could potentially cause the

[‡‡‡‡] For example, see Anderson WF, Gene therapy[88], in which he states: "If one were to believe the news media, gene therapy is both a scientific failure and unsafe. Is this gloomy picture true? Fortunately, no. The paper by Cavazzana-Calvo *et al.* (1) on page 669 of this issue provides an example of the exciting results that are starting to be obtained in human gene therapy clinical trials."[89]

sort of cellular defects that led to cancer in the SCID patients."[107] Despite several qualifying statements and detail about the study,[§§§§] a lead investigator in the study denounced the article as "suggest[ing] that there is a reasonable probability that recombinant AAV vectors may cause or contribute to cancer in gene therapy subjects."[108] His assertions were backed by the presidents of the American and European Societies of Gene Therapy, who wrote to the editors of *Nature* to complain that the article "present[ed] an entirely negative view of the field of gene therapy," and then published their missive in the latter's journal, the *Journal of Gene Medicine*.[109] Another example where major figures stepped in to neutralize what seemed like balanced reports was after the publication of two gene-transfer trials in the *New England Journal of Medicine* in 1996. Responding to a commentary that stated that the trials "underscor[ed] some of the difficulties involved in translating gene therapy from the bench to the bedside,"[110] two cystic fibrosis researchers complained that "the two articles taken together may send an unbalanced message ... it would be a great pity if they pushed the pendulum too far [away from over-enthusiasm] by ignoring the possibilities for nonviral delivery systems for gene therapy."[111][¶¶¶¶] Nelkin describes another example from a different translational research realm. When the *New York Times* ran a cautionary article about the promise of interferon, researchers reproached the reporter for expressing doubt, warning that it could affect their funding prospects.[112]

Morale boosting within the field: This strategy was, for the most part, sheltered from public view. One example was expressions of exasperation with the media and public. A leading researcher lamented that the cure of six children in the Milan ADA-SCID trial had "been largely ignored in 'bad news' stories of gene therapy." Asked about the most challenging aspect of being a leader in the field, the commentator answers "it's imparting a sense of the field's momentum to outside people such as foundation officers and NIH directors, who often remain unaware of our progress after having been saturated with negative media coverage of gene therapy. Poor public perception works against us, translating into reluctant funding agencies and clinical trial participants."[113]

Appraising hard times

All are reasonable responses of various clinical champions, soldiering against various regulatory, technical, circumstantial, and social obstacles. As with great expectations, the claims embedded in hard times discourses are not unreasonable. Nevertheless, they embody positions, definitions, and assumptions that deserve closer analysis.

The strategy of historical precedent asserts that the problems of gene transfer are problems faced by all novel fields. It presents the public – and fellow researchers – with a naturalistic view of the field's setbacks, and in so doing, seems to implicitly exonerate human commissions and omissions that have checked the field's progress. True, medical progress is always a slog, and

§§§§ The article states: "... researchers caution that the vector used in the SCID trials, which was based on a retrovirus, is very different from the adeno-associated vector," quotes a scientist saying "Adeno-associated vectors clearly have a better safety profile than retroviral vectors," and concludes with a statement from an AAV trialist: "I don't think we need to modify anything at this point ... But this is a risk we'll have to address before the vector is in widespread use."[107]

¶¶¶¶ One venue where I have personally experienced defensiveness in its most unvarnished form is in peer review. In particular, to even put the phrase "remote risk of malignancy" in a paper on the ethics of hemophilia gene transfer is to invite a firestorm from certain researchers. That this kind of outrage is expressed only behind the protective costume of anonymous peer review is yet another example of dramaturgical practice in public communications.

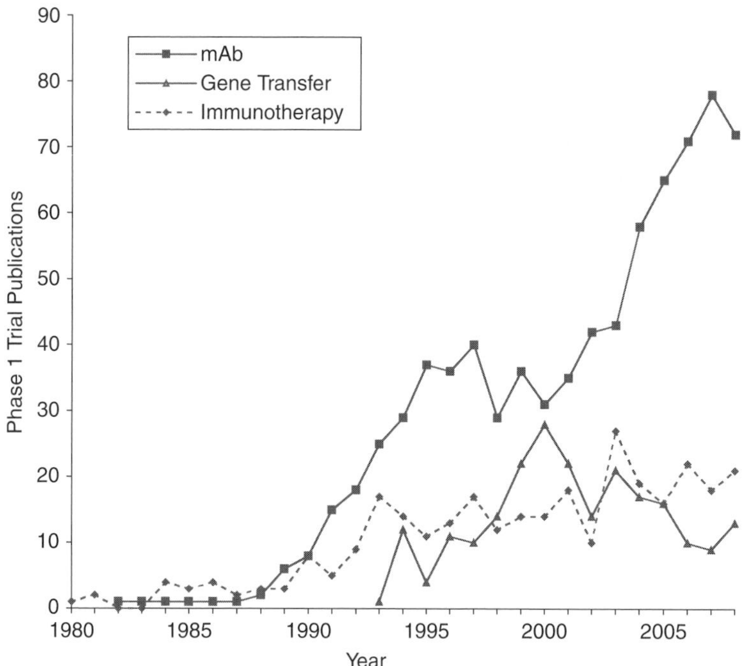

Figure 9.1 Volume of phase 1 trials published in 3 translational research areas. This graph mainly shows trends in published phase 1 trials; numeric estimates for a given year should be interpreted with caution, as some trials might be missed in search because they do not describe themselves as "phase 1." Methods: *PubMed* was searched on October 6, 2008 as follows: Monoclonal antibodies (mAb): monoclonal antibod* AND (phase 1 OR phase I) NOT phase II; Gene transfer: "gene therapy" OR "gene transfer" AND (phase 1 OR phase I) NOT phase II; Immunotherapy (cancer): "immunotherapy" AND (cancer OR oncology) AND (phase 1 OR phase I); all searches were limited to clinical trials article type. Data for 2008 were projected by dividing 2008 trial volume by 0.75.

members of the public should not expect immediate medical breakthroughs. But as has been argued throughout this book, there are also important ways in which problems encountered in gene transfer are of a different sort than those encountered in the past. The development of bone marrow transplantation, for instance, predated the era of Bayh-Dole, and decisions to initiate human testing were never as heavily driven by commercial pressures as they were for gene transfer in the early 1990s. The development of treatments for cancer and childhood leukemia involved numerous innovative social strategies, including the formation of cooperative research groups and the standardization of various protocol features.[114,115] Though gene-transfer researchers have evolved their own strategies for organizing and coordinating effort (see Chapter 4), many early efforts in the field were marked by their fragmentation and rivalry. The social and political milieu in which monoclonal antibodies were developed more closely resembles that for gene transfer. While their development involved punctuated progress and deaths occurred in studies, setbacks were not nearly as profound.***** Thus, whereas the volume of published phase 1 human gene-transfer trials declined from its peak in 2000, the volume for monoclonal antibodies declined only modestly – and relatively briefly (see Figure 9.1). Above all, expectations surrounding these earlier technologies were probably not as high as they were for gene transfer. If the gene-transfer community draws solace from a

***** Deaths occurred in trials of adult cancer, not children or healthy volunteers, and they occurred at later stage in development.[116]

history that naturalizes the field's challenges, researchers in stem-cell therapies have looked to the history to opposite effect, using gene transfer to warn against blunder.†††††

The other three strategies beg certain questions and perhaps reveal deficiencies in expectation management. For example, if major toxicities are a predictable outcome of powerful medical technologies, why were adversities encountered during the development of gene transfer so unexpected? Similarly, gene-transfer researchers have a right, if not an obligation, to correct the record when they feel that their research has been described inaccurately. But this works both ways: when journalists and other scientists project expectations that are inconsistent with consensus opinion, unless researchers correct this as well, they risk undermining their credibility as neutral scientific authorities. Moreover, the certainty with which many defensive positions are asserted only reinforces an impression of arrogance.

As for victimhood, while it might be true that public and private investment has retreated from gene-transfer clinical research, there's no evidence to support the suggestion that press coverage of the field is unremittingly negative,‡‡‡‡‡ or that expectation is currently undersubscribed. It is true that reports almost invariably describe the field as embattled, and struggling to live up to its early promise. And it's also true that many scientists working in related fields perceive that the field as "moribund" or "dead in the water."§§§§§ But public polling repeatedly

††††† Quoting Chuck Murry, of the University of Washington: "Given that nobody really knows what is going on, Murry is not alone in thinking that it was unwise to start clinical trials so soon. No serious safety problems have so far arisen in the cardiac trials, but stem cell biologists worry that rushing into the clinic on the basis of unreplicated findings could end in disaster. Their nightmare scenario is a repeat of the 1999 Jesse Gelsinger tragedy, in which an 18-year-old volunteer with a liver disease died from an inflammatory reaction to the virus used to deliver genes. The case cast a pall over the field of gene therapy that remains to this day."[117] Another quote from a famous researcher: "... there is a vivid precedent that illustrates the dangers of over-exuberance, namely, the field of gene therapy ... From the outset, in my view, scientists involved in gene therapy research underestimated the technical difficulties as well as the date when viable therapies would be available. They overestimated the numbers of diseases that would benefit from gene therapy, raising the expectations of suffering patients. Clinical experiments were undertaken that had no good foundation in basic research, and the experiments were so anecdotal that no reliable conclusions could be drawn ... The next blow to the reputation of this field came in 1999 when a young man named Jesse Gelsinger died in a gene therapy feasibility study at the University of Pennsylvania. This event discredited not just the researchers involved, but the field as a whole. A black cloud still hangs over it, discouraging bright young people from studying gene therapy and putting everyone, the serious scientist and the dilettante alike, under heightened scrutiny. As of today, 25 years after the beginning of gene therapy research, the FDA has yet to approve a single protocol. This is not a scenario that anyone would wish on a new and exciting field."[118]One news story said that stem cell biologist Evan Snyder "believes more work must be done before doctors try the experiment on humans ... If something goes awry, opponents of killing embryos for research will be poised to quash future research." Snyder was later quoted as saying, "The last thing we need is another Jesse Gelsinger."[119]

Finally *Science* reported: "many scientists worry that Geron is moving too fast. They point to gene-therapy trials in which one young patient died of an unexpected immune reaction and others developed deadly leukemia ... Gene therapy is a paradigm that we can learn from," says neuroscientist Douglas Kerr of Johns Hopkins University in Baltimore, MD. "They have actually induced harm in patients, and that has set back the field."[120]

‡‡‡‡‡ While not specifically directed toward gene transfer, there have been quantitative studies of biotechnology news coverage. These generally favor coverage of medical applications of biotechnology, especially in North America. For examples, see: Nisbet & Lewenstein[121]; Gaskell & Bauer[122]; and Bauer.[123]

§§§§§ These are actual quotes from two well-regarded researchers, one a cancer molecular biologist, another an infectious-disease researcher. Because they were made in conversation, I am not providing attribution.

demonstrates widespread receptivity for the use of gene transfer against human illnesses. For example, one large UK survey showed that 82 percent of respondents favor the use of somatic gene transfer for conditions like heart disease; 76 percent believed that genetic treatments would significantly reduce human suffering.[124] In the USA, a *Wall Street Journal*/Harris Interactive Poll found that 87 percent strongly or somewhat support the use of gene transfer to treat people who have or are likely to develop a particular disease.[125] And a scan of news databases reveals numerous stories in mainstream outlets charting the field's progress. Thus, nine months after the Penn OTC debacle, the *New York Times* runs a front page story headlined "Despite ferment, gene therapy progresses;"[126] the *Wall Street Journal*'s story is titled "Virus revives hopes for gene therapy – Primitive AAV shows promise as a safer means to carry healthy genes into body."[127] A 2001 story by Rick Weiss in the *Washington Post* begins with the lead "Human cells genetically engineered to produce a blood-clotting factor can be transplanted safely into people with severe hemophilia and can reduce the number of bleeding episodes in those patients, researchers reported yesterday."[128] A story the following year in London's *Sunday Telegraph* reports on the same study: "transplants into muscle have cut the dependence of eight haemophiliacs on the expensive [Factor IX] injections."[129] More recently, the *Wall Street Journal* ran a story on a study that reported an objective tumor regression in two of 17 volunteers. The lead states "for the first time in the history of cancer treatment, gene therapy has apparently succeeded in shrinking and even eradicating large, metastatic tumors."[130] The popular news show *Good Morning America* presents a breathless interview with Parkinson's disease researcher Michael Kaplitt and study volunteer Nathan Klein; "your heart leaps that this is the answer, that this is hope" says host Diane Sawyer.[131]

A more troubling feature with some victimhood discourses is that it seems to denigrate its audience by presuming that there is only one, reasonable view of the field's potential, which researchers alone have sufficient authority to recognize. It might be unreasonable to view gene transfer as a monolithic failure, but surely reasonable people can view novel research arenas like gene transfer with caution or skepticism.

Nevertheless, the strategies are also consistent with what was argued under great expectations: in the face of setbacks, clinical champions tend to "accentuate the positive," and when their research runs aground, they must find ways to defend their research programs. One way of doing this is by marginalizing critics.[¶¶¶¶¶]

The consequences of expectation mismanagement

One might question the parallels between gene transfer and bone marrow transplantation, or recoil at occasional displays of defensiveness. But why should the strategies of hard times matter ethically? Why should any form of expectation management matter ethically?

The ethics and policy literatures generally describe three potential adverse outcomes from overextending expectations: first, they harm patients by creating "false hopes." "False hopes" can have various adverse consequences: they can lead persons to discount warnings, to postpone important decisions, or to take actions that have major negative consequences. Once a hope is exposed for its "falsity," it can cause disappointment or embarrassment. Second, overextended expectations can frustrate patient autonomy by distorting the understanding of risk and benefit. This would lead to invalid informed consent. Such invalid informed consent

---

[¶¶¶¶¶] Goffman wrote that "backstage derogation of the audience serves to maintain the morale of the team. And when the audience is present, considerate treatment of them is necessary, not for their sake... merely, but so that continuance of peaceful and orderly interaction will be assured.[132]

might be tolerable in the context of medical care, because at least there is greater assurance that the objectives of patient and physician are aligned. In the context of clinical research, however, this kind of misinformed consent is far more troubling, because the volunteer's objective on entering the trial is often therapeutic, and thus in tension with that of the clinical investigator. Third, overextended expectations might undermine trust in medical research.[133] Patient communities might become wary of research fields; policy-makers might withdraw research funds or impose cumbersome regulations, and prospective volunteers might be less inclined to consider enrollment in trials.

The logical case against expectation mismanagement is as strong as that against various forms of deception, and all three outcomes are plausible and at least supported by anecdote. Nevertheless, an ethics of expectation management faces several complexities.

For example, what, precisely, is meant by "false hope"? There is often an implicit, naive realism in this concept: a person with a 0.1 percent chance of surviving a terminal illness may have false beliefs if he or she believes that the odds of survival are 50 percent. However, there is nothing unrealistic in *hoping* that he or she will beat dismal odds. Nor is the patient demonstrably incorrect for believing that, whether because of divine intervention, healing crystals, or chance, he or she holds the winning lottery ticket.****** Further, some commentators note that hope often "resists speculating" about probabilities.[138]

The relationship between expectation and consent is also subtle, and ethicists and policy-makers should be careful about infantilizing patients and the public by assuming that each represents uncritical and passive receptacles of expectation management practices. The angiogenesis inhibitor episode described earlier in this chapter provided the occasion for a series of surveys showing that cancer patients in phase 1 studies often engage critically with information about emerging technologies. One found that patients referred to trials of angiogenesis inhibitors who learned about the study through media reports had a greater understanding of the study's purposes, and were no more motivated by the prospect of medical benefit than persons who learned about the trial through other channels.[139] Another found that cancer patients often regard news accounts of treatment breakthroughs with skepticism.[140] Similar skepticism about medical breakthroughs has been corroborated in other studies as well.[76] A different study performed in Italy after aggressive media campaigns promoting an alternative, unproven cancer therapy found that news coverage of the treatment increased the feelings of hope for cancer patients and convinced them of the treatment's effectiveness.[141] But only a small percentage of patients said they would be willing to try the therapy, and the media campaign did not seem to measurably influence patient–physician decision-making.[142]

With respect to the suggestion that sensationalism undermines trust, despite occasional scandals and overzealous media coverage in biomedical research, both the scientific and medical communities continue to rank among the most trusted professions,†††††† and public opinion continues to stand strongly behind gene transfer.[124, 125] Still, were it true that sensationalism undermined public confidence in science, would this necessarily be regrettable? Many scientists might answer in the affirmative. But science is often said to be a form of

****** For a series of papers exploring the concept of hope and false hope, see: Penson *et al.*[134], Ruddick[135] and Simpson[136]. One productive place to begin a further analysis of false hope would be through an examination of the psychology literature on positive psychology and "protective pessimism." See, for instance, Norem & Chang.[137]

†††††† For example, a 2006 Harris Interactive Poll found doctors and scientists ranked first and third, respectively, in response to the question "would you generally trust each of the following types of people to tell the truth, or not?" Actors and lawyers ranked the lowest.[143]

"organized skepticism."[144] Why should the critical outlook that science inculcates in its trainees become undesirable when taken up by members of the public – or for that matter, by ethicists?

Great expectations represents only one mode of expectation management. In what ways might the discourses described in hard times be ethically consequential? Here, the concern is that such strategies interrupt opportunities for serious engagement between expectant publics and researchers. For example, comparisons with historical examples would not be necessary had public expectations not outrun consensus scientific opinion. Researchers might ask themselves, and each other, how expectations of gene transfer escalated so quickly, and why researchers were not effective in dampening these expectations. Denigrating public understandings or dismissing risk perceptions further hardens a boundary between researchers and members of the public. This makes serious dialog and reflection all the more difficult to achieve.

The ethical management of expectations

Though expectations attend to many kinds of clinical research, they are particularly volatile where interventions are highly novel – where, in essence, a commitment to pursue research is driven less by stable knowledge than by imagination. The lack of success in optimally managing expectations may represent yet another way in which the field of gene transfer lost its way in translation. It is encouraging that media relations have made it into a training module for researchers at the US National Cancer Institute,[145] and ethics policies of India,[61] and the American College of Physicians.[60] Nevertheless, the limitations of current discourses on "sensationalism" and expectation management are yet another way in which the ethics of early-phase research is not adequately developed.

Some might question whether research teams *should* manage expectations in the first place. "Expectation management" might seem susceptible to instrumentalizing publics, and medical research teams might use such practices to advance parochial objectives. Moreover, "expectation management" suggests an asymmetrical relationship between medical research teams (agents) and various publics (who, as objects, are implicitly passive and silent). According to this view, the public should not be regarded as a passive entity, and should at least be granted some role in shaping researchers' expectations.

However, concerns about instrumentalizing take a cynical view of expectation management. There is no reason why expectation management cannot advance mutual objectives, or reconcile public expectation with consensus scientific opinion, or improve understanding, or prevent undesirable outcomes. Who could possibly object to the expectation management practices surrounding HIV vaccine trials to prevent volunteers from engaging in riskier behavior because of expectation about vaccine efficacy?[146] Or those performed in order to assure sustained public support for politically sensitive research?[147,148] Whether expectation management, on balance, leads to abuses is largely an empirical question. Nevertheless, given the possible benefits described above, the appropriate response to abuse would be to determine what kinds of circumstance lead to abuses, and to develop policies or structures that prevent such abuses.

Concerns about asymmetries of communication depend critically on the way expectation management is practiced. Some asymmetries cannot be eliminated, since clinical and preclinical research is an esoteric realm, and untutored members of the public will generally lack the sophistication to evaluate a study's scientific and clinical value. Moreover, as the producers of science, research teams inevitably initiate (and hence, frame) communications.

Table 9.1 Expectation management in translational research: recommendations for researchers, sponsors, and host institutions

(1) Develop and maintain closer relations with interested publics (e.g. patient advocates, policy-makers, members of the public)
(2) Respond to, and if necessary, counter "Great Expectations" communication mode
(3) Third party, critical review of press communications
(4) Communications should be attentive not merely to language, but also to gesture and symbolism

But prescriptions for expectation management that move beyond calls for greater accuracy and sobriety can have a more reciprocal relationship between publics and scientists. Throughout this chapter, I have characterized expectation management as a sort of performance, with researchers attempting to direct public attention and shape perceptions. The core policy I would urge is an expectation management process that is relatively transparent and that fosters equilibrium of expectation among research teams, members of the public, and the biomedical peer community.[‡‡‡‡‡‡]

This might take several forms (Table 9.1). First, research teams might cultivate closer relations with disease advocates or community advisory boards, perhaps through regular meetings or conference calls.[§§§§§§] These should be organized not merely toward strategic ends, but also for monitoring patient communities' expectations and making the dreadful tedium and contention in normal scientific advance more manifest.

Second, peer researchers and leadership should be as energetic about countering great expectations as they are in promoting displays of hard times. Though it might not be apparent to outside observers, gene transfer has a rich tradition of self-criticism and internal dissent (recall Box 9.1).[151] And scientists outside the field have often spoken truth to underpowered data: "I don't want to argue that gene therapy will never be achieved. I just want to dampen this near hysteria that we have solved the problems of genetic disease. We have not."[152] This controversy and dissent should be more visible, and fostered by professional societies and journal editors.

Third, institutions, research teams, and funding agencies should subject their press releases to some sort of peer review, just as recruitment materials for clinical trials are typically vetted by IRBs.

Fourth, review structures should hold research teams accountable not merely for their language, but also for their symbolism and gesture. In particular, they should discourage investigators or patient advocacy groups from presenting studies as therapeutic exercises. They might press teams to disclose scientific contingencies. They might also discourage the use of patient-subjects to display or justify the initiation of trials. Throughout this book, I have argued that translational researchers should favor a scientific model over a regulatory model of early-phase studies; that first-in-human studies are part of an iterative process whereby novel interventions are characterized and refined. Entry into human testing should thus not be presented as a harbinger of clinical application.

---

‡‡‡‡‡‡ For a discussion of more dynamic models of public relations that has relevance for expectation management, see Leeper.[149]

§§§§§§ For example, Paul Kaplan, the general manager of a joint venture developing an enzyme replacement therapy for Pompe's disease, holds monthly "progress report" conference calls with disease advocates.[150]

Seeing the light or blind faith?

And what about Lancelot? The use of this photogenic animal is somewhat troubling. Though he may be a dog, Lancelot's presentation appeals strongly to therapeutic and patient narratives; his appearance may have been instrumental in shifting sentiment among regulators and the RAC in favor of a first-in-human protocol that is ethically contentious for its inclusion of children as young as eight (normally, translational studies for a non-fatal disease would be conducted first in adult volunteers). Yet in both public and semi-private forums, Jean Bennett is circumspect about the public reception of her work, and as websites maintained by groups like the Foundation for Fighting Blindness and the Foundation for Retina Research attest, she is deeply engaged with the patient communities that have much to lose from mismanaged expectations.

Whatever expectation these preclinical studies might have generated pales in comparison with preliminary reports from the first human studies. In two back-to-back studies published in the *New England Journal of Medicine*, researchers at the Scheie Eye Institute and Moorfields Eye Hospital in London reported on the first three patients receiving AAV vectors for LCA. Both studies showed some improvements in physiological parameters; the former showed "modest improvement in measures of retinal function on subjective tests of visual acuity"[153], while the latter showed "no clinically significant change in visual acuity."[154] The reports were greeted with a deluge of expectation: *Washington Post* reporter Rick Weiss called the news "something short of miraculous;"[155] patients were put before Diane Sawyer ("this is really the curtains parting on her future"[156]); in a third trial, a patient described the "overwhelming moment" when he saw the Florida sunshine.[157] And disease advocates called the report a "dramatic breakthrough."[158]

Expectation around LCA gene transfer is now at a fever pitch. Imagine the response should, in the coming years, volunteers develop complications or promising clinical effects deteriorate. Time will tell whether the Pennsylvania team – along with the field of gene transfer and patient advocates – managed expectations effectively.

References

1. Acland GM, Aguirre GD, Ray J, *et al.* Gene therapy restores vision in a canine model of childhood blindness. *Nat Genet* 2001; **28**(1): 92–5.
2. Weiss R. Gene treatment restores vision in blind dogs; Study offers humans hope. *Washington Post* 2001 April 28.
3. Smith S. Were blind, but now dogs see. *The Penn Current* 2001 May 31.
4. Giresi JC. In dogged pursuit of ocular health. *The Pennsylvania Gazette* 2001 March 5.
5. Anonymous. Researcher calls for more federal funding for eye research. *Cornell Chronicle* 2001 June 14.
6. The Foundation Fighting Blindess. *2001 Annual Report – A Cure is in Sight*, 2001.
7. Onion A. How dog genes may help the blind see: how humans' best friend could also become their savior. *ABC News*, 2003 October 13.
8. Couzin J. Clinical research. Advocating, the clinical way. *Science* 2005; **308**(5724): 940–2.
9. Kittredge C. Under the microscope again. *The Scientist* 2005; **19**(9): 15.
10. American Society of Gene Therapy. Special Media Program – "Horizons in Gene Therapy". *Proceedings of the ASGT 10th Annual Meeting*, 2007 May 30 to June 3. Seattle, WA, ASGT, 2007.
11. Gura T. Hemophilia. After a setback, gene therapy progresses... gingerly. *Science* 2001; **291**(5509): 1692–7.
12. Nelkin D. *Selling Science: How the Press Covers Science and Technology*. New York, W. H. Freeman Co., 1995.
13. Friedman SM, Dunwoody S, Rogers CL, eds. *Communicating Uncertainty: Media*

Coverage of New and Controversial Science. Mahwah, NJ, Lawrence Earlbaum Associates Publishers, 1999.

14. Dresser R. Advocacy for Accuracy in Reporting. In: *When Science Offers Salvation: Patient Advocacy and Research Ethics*, Chapter 7. New York, Oxford University Press, 2002; 129–50.
15. Phillips J, Beckman F. Good relations – PR for the biotech business. *Nat Biotechnol* 2001; **19**: BE34–6.
16. Goffman E. *The Presentation of Self in Everyday Life*. New York, NY, Doubleday, 1959.
17. Hilgartner S. *Science on Stage – Expert Advise as Public Drama*. Stanford, CA, Stanford University Press, 2000.
18. Brauerhoch FO, Ewen C, Sinemus K. Talking biotech with the public. *Biotechnol J* 2007; **2**(9): 1076–80.
19. Advisory Committee on Human Radiation Experiments. *Final Report: Chapter 16: Subject Interview Study*, ACHRE Report, 1995.
20. Brown N, Michael M. A sociology of expectations: Retrospecting prospects and prospecting retrospects. *Technology Analysis and Strategic Management* 2003;**15**(1): 3–18.
21. Hedgecoe A. *The Politics of Personalized Medicine: Pharmacogenetics in the Clinic*. New York, NY, Cambridge University Press, 2004.
22. Rosenberg LE, Schechter AN. Gene therapist, heal thyself. *Science* 2000; **287**(5459): 1751.
23. Thompson L. At age 2, gene therapy enters a growth phase. *Science* 1992; **258**(5083): 744–6.
24. Friedmann T. Rigor in gene therapy studies. *Gene Ther* 1995; **2**(6): 355–6.
25. Arkin LM, Sondhi D, Worgall S, Suh LH, Hackett NR, Kaminsky SM, *et al.* Confronting the issues of therapeutic misconception, enrollment decisions, and personal motives in genetic medicine-based clinical research studies for fatal disorders. *Hum Gene Ther* 2005; **16**(9): 1028–36
26. Chapters 5–7 of Pieters T, *Interferon: The Science and Selling of a Miracle Drug*. New York, Routledge, 2005).
27. Cohen, J. The confusing mix of hype and hope. *Science* 2002; **295**: 1026.
28. Anderson WF. Musings on the struggle – Part III: The October 3 RAC meeting. *Hum Gene Ther* 1993; **4**(4): 401–2.
29. Recombinant DNA Advisory Committee (RAC). Minutes of Meeting, October 7–8 1991.
30. A one-in-a-million worst case. *Time* March 26, 1984.
31. UPI. Young Texas transplant patient is among many aided by Nobel researcher. *The New York Times* 1985 October 15.
32. Anonymous. Her destiny? Dying young, until MDs altered genes. *Winnipeg Free Press* 1994 April 2; A5.
33. Rubin R. Gene therapy targets defect. Technique could replace transplants for patients like Stormie Jones. *Dallas Morning News* 1991 October 21; 6D
34. Angier N. A new gene therapy to fight cholesterol is being prepared. *New York Times* 1991 October 29; A3.
35. Gorner P, Kotulak R. Gene therapy aimed at deadly cholesterol. *Chicago Tribune* 1992 February 19; A5.
36. University of Pennsylvania Medical Center announces first human gene therapy institute in the world. *PR Newswire*, 1992 December 7.
37. Grossman M, Raper SE, Kozarsky K, Stein EA, Engelhardt JF, Muller D, *et al.* Successful ex vivo gene therapy directed to liver in a patient with familial hypercholesterolaemia. *Nat Genet* 1994; **6**(4): 335–41.
38. Weatherall D. Heroic gene surgery. *Nat Genet* 1994; **6**(4): 325–6.
39. Angier N. Gene experiment to reverse inherited disease is working. *The New York Times* 1994 April 1.
40. Weiss R. Human gene therapy achieves a milestone. *The Washington Post* 1994 April 1.
41. Ritter M. Gene therapy lowers woman's lethal cholesterol levels, study says. *Associated Press*, 1994 March 31.
42. Williams RS. Human gene therapy – of tortoises and hares. *Nat Med* 1995; **1**(11): 1137–8.
43. Angier N. Gene therapy helps cure inherited disease. *Pittsburgh Post-Gazette* 1994 April 1.
44. Casey J. The next wonder drug may not be a drug: it may be a therapeutic gene – and

results are improving. *Business Week*, 1994: 84.
45. Weiss R. Human gene therapy achieves a milestone. *The Washington Post* 1994 April 1.
46. Quoting Kenneth Culver; Genes implanted in liver help lower cholesterol. *St. Petersburg Times* 1994 April 1.
47. Rensberger B. Scientists close in on cystic fibrosis cure. *Washington Post* 1992 January 10.
48. Detjen J. Gene therapy may be key to curing lung disease. *LA Daily News* 1991 April 19.
49. Congressional testimony 10 February 1992. Before the Subcomittee on Energy, US House of Representatives Committee on Science, Space and Technology. *Human Radiation Experimentation, Ethics, and Gene Therapy*.
50. Angier N. Panel permits use of genes in treating cystic fibrosis. *The New York Times* 1992 December 4.
51. Recer P. Experimental gene therapy begun for cystic fibrosis. *Associated Press*, 1993 April 19.
52. Herman R. Gene therapy OK'd for cystic fibrosis – approval gives hope to 30,000 Americans. *Austin American-Statesman* 1992 December 4.
53. Herman R. Gene therapy to be tested in cystic fibrosis patients; Virus to be used to correct lung-cell defects. *The Washington Post* 1992 December 4.
54. Gene therapy OK'd for cystic fibrosis – Scientific panel gives unanimous approval. *The San Francisco Chronicle* 1992 December 4.
55. Angier N. Gene therapy begins for fatal lung disease. *The New York Times* 1993 April 20.
56. Knox RA. Cystic fibrosis is reported corrected by gene therapy. *The Boston Globe* 1993 October 15.
57. Shuchman M, Wilkes MS. Medical scientists and health news reporting: A case of miscommunication. *Ann Intern Med* 1997; **126**(12): 976–82.
58. Nelkin D. *Selling Science: How the Press Covers Science and Technology*. New York, W. H. Freeman Co., 1995; 170.
59. Dresser R. *When Science Offers Salvation: Patient Advocacy and Research Ethics*. New York, NY, Oxford University Press, 2002; 139.
60. Snyder L, Leffler C. Ethics manual: fifth edition. *Ann Intern Med* 2005; **142**(7): 560–82; 578.
61. Indian Council of Medical Research. *Ethical Guidelines for Biomedical Research on Human Participants*. New Delhi, Royal Offset Printers, 2006; 32.
62. Quoted in Marshall E. Gene therapy's growing pains. *Science* 1995; **269**(5227): 1050–5.
63. Fox RC, Swazey JP. *The Courage to Fail: A Social View of Organ Transplants and Dialysis*, new edn. Chicago, University of Chicago Press, 2001.
64. Connor S. Hopes rise of therapy for cystic fibrosis. *The Independent* 1993 October 16; 5.
65. Angier N. Cystic fibrosis experiment hits a snag. *The New York Times* 1993 September 22; 12.
66. Brown D. Medicine: Gene therapy for cystic fibrosis. *The Washington Post* 1993 March 22; A02.
67. Kolata GA. A cautious awe greets drugs that eradicate tumors in mice. *The New York Times* 1998 May 3.
68. Marshall E. Cancer: The roadblocks to angiogenesis blockers. *Science* 1998; **280**(5366): 997.
69. Ryan DP, Penson RT, Ahmed S, Chabner BA, Lynch TJ, Jr. Reality testing in cancer treatment: the phase I trial of endostatin. *Oncologist* 1999; **4**(6): 501–8.
70. Schwitzer G. A statement of principles for health care journalists. *Am J Bioeth* 2004; **4**(4): W9–13.
71. McIlroy A. 'How far would you go for someone you love?': To save his son, John Davidson walked 11 572 kilometers and raised $9.5 million, Anne McIlroy reports. Why, then, is he so sad? *The Globe and Mail* 2007 June 16.
72. This position is expressed in: Condit CM. How geneticists can help reporters to get their story right. *Nat Rev Genet* 2007; **8**(10): 815–20.
73. Fox RC, Swazey JP. *The Courage to Fail: A Social View of Organ Transplants and Dialysis*, new edn. Chicago, Ill, University Chicago Press, 2001; 399–400.
74. "Cancer researcher: 'This is just a start'" CNN August 31, 2006. Available at: http://edition.cnn.com/2006/HEALTH/08/31/

cnna.rosenberg/index.html; last accessed December 5, 2007.

75. Nelkin D. *Selling Science: How the Press Covers Science and Technology*. New York, W. H. Freeman Co., 1995; at 73.

76. Brown N. Hope against hype – Accountability in biopasts, presents and futures. *Science Studies* 2003; **16**(2): 3–21.

77. King NM. Experimental treatment. Oxymoron or aspiration? *Hastings Cent Rep* 1995; **25**(4): 6–15.

78. Weiss R. Getting new genes; A radical treatment for a fatal liver flaw. *Washington Post* 1994 February 15.

79. Crawford D. OSU medical center begins first human gene therapy trial. The Ohio State University Medical Center, *Press Releases*, 1999.

80. Dubowitz V. Therapeutic efforts in Duchenne muscular dystrophy; the need for a common language between basic scientists and clinicians. *Neuromuscul Disord* 2004; **14**(8–9): 451–5.

81. Press Release, OSU Medical Center Begins First Human Gene Therapy Trial. 1999 September 3. available at: http://medicalcenter.osu.edu/mediaroom/press/article.cfm?ID=847. Last accessed November 28, 2007.

82. Brown N, Michael M. A sociology of expectations: Retrospecting prospects and prospecting retrospects. *Technology Analysis and Strategic Management* 2003; **15**(1):3–18.

83. Brennan AL, Geddes DM, Gyi KM, Baker EH. Clinical importance of cystic fibrosis-related diabetes. *J Cyst Fibros* 2004; **3**(4): 209–22.

84. Rogers MH, Lwin R, Fairbanks L, Gerritsen B, Gaspar HB. Cognitive and behavioral abnormalities in adenosine deaminase deficient severe combined immunodeficiency. *J Pediatr* 2001; **139**(1): 44–50.

85. Lang AE, Obeso JA. Challenges in Parkinson's disease: restoration of the nigrostriatal dopamine system is not enough. *Lancet Neurol* 2004; **3**(5): 309–16.

86. Orkin SH, Motulsky AG. *Report and Recommendations of the Panel to Asses the NIH Investment in Research on Gene Therapy*. US National Institutes of Health, Office of Biotechnology Activities, 1995.

87. Stephenson J. Gene therapy trials show clinical efficacy. *JAMA* 2000; **283**(5): 589–90

88. Anderson WF. Gene therapy:the best of times, the worst of times. *Science* 2000; **288**(5466): 627–9.

89. Cavazzana-Calvo M, Hacein-Bey S, de Saint Basile G, Gross F, Yvon E, Nusbaum P, *et al*. Gene therapy of human severe combined immunodeficiency (SCID)-X1 disease. *Science* 2000; **288**(5466): 669–72.

90. Judson HF. *The Glimmering Promise of Gene Therapy*. MIT Technology Review November 1, 2006.

91. Manno CS, Pierce GF, Arruda VR, Glader B, Ragni M, Rasko JJ, *et al*. Successful transduction of liver in hemophilia by AAV-Factor IX and limitations imposed by the host immune response. *Nat Med* 2006; **12**(3): 342–7.

92. Urbanowicz N. Genzyme acquires Avigen's gene therapy technology. *Dowjones Newswires*, 2005 December 21.

93. Howard K. First Parkinson gene therapy trial launches. *Nat Biotechnol* 2003; **21**(10): 1117–18.

94. www.widmeyer.com/general/widmeyer_announ_1.asp

95. Sarasohn J. Google bolsters its Washington presence. *The Washington Post* 2006 May 25.

96. Advances in gene therapy for hemophilia; lessons for similar solutions for other diseases. *PR Newswire* 2002 February 28.

97. Feldbaum CB. Gene therapy's remedy: Setbacks should not prevent use. *The Washington Times* 2000 January 26.

98. High KA. Stakeholders' conference sharpens focus on challenges of clinical gene transfer. *Mol Ther* 2005; **12**(4): 581–2.

99. Harris G. Gene therapy is facing a crucial hearing. *The New York Times* 2005 March 3.

100. Quote from Donald Kohn in: Harris G. Gene therapy is facing a crucial hearing. *The New York Times* 2005 March 3.

101. Smith C. Curing disease from inside the cell; 50 years after DNA breakthrough, Seattle is a leader in gene research. *Seattle Post-Intelligencer* 2006 February 28; at A1.

102. Friedmann, T. Presentation. *NIH (RAC) Gene Transfer Safety Symposium: Current Perspectives on Gene Transfer for X-SCID*. Bethesda, MD, March 15, 2005. Available

at: www.webconferences.com/nihoba/RAC%20keynote_Friedman.pdf. Accessed July 4, 2007.

103. Friedmann T. Keynote Speech. *NIH (RAC) Gene Transfer Safety Symposium: Current Perspectives on Gene Transfer for X-SCID.* Bethesda, MD available at: www.webconferences.com/nihoba/RAC%20keynote_Friedman.pdf; 2005.
104. Hall, Stephen S. A death in Philadelphia. *Technology Review* 2002 February 21.
105. Check E. A tragic setback. *Nature* 2002; **420**(6912): 116–18.
106. Weiss R. Dream unmet 50 years after DNA milestone; gene therapy debacle casts pall on field. *The Washington Post* 2003 February 28.
107. Check E. Harmful potential of viral vectors fuels doubts over gene therapy. *Nature* 2003; **423**(6940): 573–4.
108. Kay MA, Nakai H. Looking into the safety of AAV vectors. *Nature* 2003; **424**(6946): 251.
109. Kohn DB, Gansbacher B. Letter to the editors of *Nature* from the American Society of Gene Therapy (ASGT) and the European Society of Gene Therapy (ESGT). *J Gene Med* 2003; **5**(7): 641.
110. Leiden JM. Gene therapy – promise, pitfalls, and prognosis. *N Engl J Med* 1995; **333**(13): 871–3.
111. Alton E, Geddes DM. Gene therapy. *N Engl J Med* 1996; **334**(5): 332; author reply 333.
112. Nelkin D. *Selling Science: How the Press Covers Science and Technology.* New York, W. H. Freeman Co., 1995; 145–146.
113. Quoting Katherine High in: Brown K. Gene therapy: still a contender. *HHMI Bulletin* 2005; **18**(2): 38–9, 63.
114. Keating P, Cambrosio A. From screening to clinical research: the cure of leukemia and the early development of the cooperative oncology groups, 1955–1966. *Bull Hist Med* 2002; **76**(2): 299–334.
115. Schilsky RL, McIntyre OR, Holland JF, Frei E 3rd. A concise history of the cancer and leukemia group B. *Clin Cancer Res* 2006; **12**(11 Pt 2): 3553s–5s.
116. Dickman S. Antibodies stage a comeback in cancer treatment. *Science* 1998; **280**(5367): 1196–7.
117. Aldhous P. Miracle postponed; In the light of the Korean scandal, many big claims about stem cells are looking decidedly doubtful. *New Scientist* 2006; **189**(2542): 42.
118. Tilghman SM. Address to the Stem Cell Institute, New Jersey, November 11, 2004. Presented at the Inaugural Symposium, the Stem Cell Institute of New Jersey.
119. Philipkoski K. Race to human stem-cell trials. *Wired* 2005 April 19.
120. Vogel G. Cell biology. Ready or not? Human ES cells head toward the clinic. *Science* 2005; **308**(5728): 1534–8.
121. Nisbet MC, Lewenstein BV. Biotechnology and the American media: the policy process and the elite press, 1970 to 1999. *Science Communication* 2002; **23**(4): 359–91.
122. Gaskell G, Bauer MW, eds. *Biotechnology 1996–2000: The Years of Controversy.* London, Science Museum Publications, 2001.
123. Bauer MW. Distinguishing red and green biotechnology: cultivation effects of the elite press. *Int J Public Opin Res* 2005; **17**(1): 63–89.
124. Wellcome Trust. *What do People Think about Gene Therapy?* London, Wellcome Trust, 2005.
125. *Public overwhelmingly supportive of genetic science and its use for a wide variety of medical, law enforcement and personal purposes.* Rochester, NY, Harris Interactive, 2006. www.harrisinteractive.com/news/allnewsbydate.asp?NewsID=1088.
126. Stolberg S. Despite ferment, gene therapy progresses. *The New York Times* 2000 June 6.
127. Langreth R. Virus revives hopes for gene therapy – Primitive AAV shows promise as a safer means to carry healthy genes into body. *The Wall Street Journal* 2000 March 29.
128. Weiss R. Cell transplants help hemophiliacs; genes engineered to produce blood-clotting factor have limited success. *The Washington Post* 2001 June 7.
129. Highfield R. Gene trials lift hope on haemophilia. *The Sunday Telegraph* 2002 February 17.

130. Schoofs M. New gene therapy appears to shrink tumors in two cases. *The Wall Street Journal* 2006 September 1.
131. New Parkinson's hope? Promising treatment helps symptoms. *Good Morning America*, 2007 June 29.
132. Goffman E. *The Presentation of Self in Everyday Life*. New York, NY, Doubleday, 1959; 175.
133. Caulfield T. Biotechnology and the popular press: hype and the selling of science. *Trends Biotechnol* 2004; **22**(7): 337–9.
134. Penson RT, Gu F, Harris S, Thiel MM, Lawton N, Fuller AF Jr, *et al*. Hope. *Oncologist* 2007; **12**(9): 1105–13.
135. Ruddick W. Hope and deception. *Bioethics* 1999; **13**(3–4): 343–57.
136. Simpson C. When hope makes us vulnerable: a discussion of patient–healthcare provider interactions in the context of hope. *Bioethics* 2004; **18**(5): 428–47.
137. Norem JK, Chang EC. The positive psychology of negative thinking. *J Clin Psychol* 2002; **58**(9): 993–1001.
138. Martin AM. Hope and exploitation. *Hastings Cent Rep* 2008; **38**(5): 49–55. According to Martin, what makes "false hope" ethically problematic is that it leaves patients and research volunteers particularly vulnerable to exploitation.
139. Pentz RD, Flamm AL, Sugarman J, Cohen MZ, Daniel Ayers G, Herbst RS, *et al*. Study of the media's potential influence on prospective research participants' understanding of and motivations for participation in a high-profile phase I trial. *J Clin Oncol* 2002; **20**(18): 3785–91.
140. Chen X, Siu LL. Impact of the media and the internet on oncology: survey of cancer patients and oncologists in Canada. *J Clin Oncol* 2001; **19**(23): 4291–7.
141. Passalacqua R, Campione F, Caminiti C, Salvagni S, Barilli A, Bella M, *et al*. Patients' opinions, feelings, and attitudes after a campaign to promote the Di Bella therapy. *Lancet* 1999; **353**(9161): 1310–14.
142. Passalacqua R, Caminiti C, Salvagni S, Barni S, Beretta GD, Carlini P, *et al*. Effects of media information on cancer patients' opinions, feelings, decision-making process and physician–patient communication. *Cancer* 2004; **100**(5): 1077–84.
143. Doctors and teachers most trusted among 22 occupations and professions: fewer adults trust the president to tell the truth. In: The Harris Poll #61: Harris Interactive; 2006. www.harrisinteractive.com/harris_poll/index.asp?PID=688
144. Merton RK. The normative structure of science (1942). In: Storer NW, ed. *The Sociology of Science*. Chicago, University of Chicago Press, 1973.
145. National Institutes of Health. *Protecting Human Research Subjects* course. National Cancer Institute.
146. See, for example: Haynes BF. Scientific and social issues of human immunodeficiency virus vaccine development. *Science* 1993; **260**(5112): 1279–86.
147. United Nations Developing Programme. *Governance of HIV/AIDS Responses: Issues and Outlook*. New York, NY, UNDP, 2006.
148. Uganda AIDS Commission, European Union, and Uganda HIV/AIDS Partnership. *Uganda Guidelines for AIDS Vaccine Research: A Guide for Vaccine Research, Development and Evaluation*. March 2006.
149. Leeper RV. Moral objectivity, Jurgen Habermas's discourse ethics, and public relations. *Public Relations Rev* 1996; **22**(2): 133–50.
150. Paul Kaplan, General Manager of the Genzyme/Pharming Joint Venture, interviewed by Kevin O'Donnell about the current state of the art regarding Genzyme's current ERT (enzyme replacement therapy) project. Transcript on file with author (www.amda-pompe.org/press.htm). Interview undated, but based on other dates in the transcript, it probably occurred in 2001.
151. Quoting Richard Mulligan: in Thompson L. Medicine's 4-year-old pioneer; first gene-therapy patient opens door to treating 4,000 inherited diseases. *The Washington Post* 1990 September 25.
152. Thompson L. Medicine's 4-year-old pioneer; first gene-therapy patient opens door to treating 4,000 inherited diseases. *The Washington Post* 1990 September 25.
153. Maguire AM, Simonelli F, Pierce EA, Pugh EN Jr, Mingozzi F, Bennicelli J, *et al*. Safety and efficacy of gene transfer for Leber's

congenital amaurosis. *N Engl J Med* 2008; **358**(21): 2240–8.

154. Bainbridge JW, Smith AJ, Barker SS, Robbie S, Henderson R, Balaggan K, *et al.* Effect of gene therapy on visual function in Leber's congenital amaurosis. *N Engl J Med* 2008; **358**(21): 2231–9.

155. Weiss R. Altered viruses reversed progressive blindness, studies say. *Washington Post* 2008 April 28.

156. *Good Morning America.* Reversing blindness: Gift of sight for young people. ABC 2008 April 28.

157. Mick H. Gene therapy shines light on blindness. *Globe and Mail* 2008 May 2.

158. Statement from Lighthouse International about the study on gene therapy concerning Leber's congenital amaurosis. *PR Newswire* 2008 April 29.

Chapter 10 Something in the sight adjusts itself: conclusions

Introduction

The patients were at the end of the line – unable to climb a flight of steps, constantly short of breath, and prepared for a last, fatal heart attack. Using a long device vaguely resembling a fishing rod, a medical team led by Texas Heart Institute cardiologist Emerson Perin navigated through a small incision in the groin to reach the patients' hearts, where he made a series of injections.

The content of the injections was bone marrow mononuclear cells – a type of adult stem cell that gives rise to different blood types – freshly pulled from the back of each patient's hip. This would be one of the first attempts at cell therapy for heart disease. The theory, grounded in an important study performed at the NIH,[1] was that some of the bone marrow would develop into cardiac cells and regenerate damaged heart tissue. Because the cells were derived from each patient, immune rejection was not anticipated to be a problem. Still, the procedure was a bold one: what if the cell mixture developed into bone? What if the cells caused abnormal rhythms in the heart? What if they clumped together and triggered an embolism?

Brazilian regulators took a favorable view of the protocol, and Perin, who was trained in Brazil and had working relationships with cardiologists there, took his study to Rio de Janeiro in 2001. The team, and their regulators, it turned out, guessed right: among the fourteen patients, none developed worrisome complications.[2] Though the study was not rigorously controlled,* volunteers receiving cells seemed to do better than patients who did not.

In communications with the media, the research team suggested that the injections "brought heart failure patients back from near-death." As one investigator put it, "these people in Brazil were reborn. The stem cell injections 'woke up' their hearts." One patient, Nelson Aguia, was described as "walking three miles around Brazil's largest soccer stadium twice a week, swimming three times a week, and working eight hours a day as a pharmaceutical sales representative – a job he gave up years ago. During Brazil's annual Carnival celebration each February, he dances in the streets."[3] Accompanying the announcement were photos of Aguia, and two senior investigators fitted with signature white jackets and nearly matched red ties. Shortly thereafter, the investigators initiated rigorously controlled, FDA-approved phase 2 studies in Texas.[4]

The challenge of uncertainty in translational clinical research

The above vignette concerns an intervention that, though not involving gene transfer, is characterized by a cluster of ethical challenges like those encountered in this book. I have offered

* Seven patients were included as controls. But because none received sham injections, they were not randomized, and all were aware that they had not received active agent, placebo effects cannot be ruled out.

three core theses about such challenges. First, novel interventions in early-phase trials present a series of ethical difficulties that, if not without precedent, are distinctive, unfamiliar, and unresolved. Second, almost two decades of experience with gene transfer provide some outline of what these challenges are and will be for similarly novel interventions. Third, experience and reflection also indicate that tools and heuristics developed for evaluating phase 3 studies have serious limitations when extended to first-in-human studies.

With respect to the first thesis, major challenges include the assessment and management of risk, appropriate subject selection, appraisal of study value, questions about when to initiate human studies (and how), ensuring fairness in trials involving economically disadvantaged communities, and regulating expectations. Much of the ethical distinctiveness represented by these challenges is traceable to the character and degree of uncertainty in translational trials. Agents themselves present a number of uncertainties, including how to describe and compare their composition,[5] how to model human response, and how to measure human response. I have also emphasized the social and institutional uncertainties as well: the roles and responsibilities of actors are often ambiguous and shifting, criteria for evaluating safety are unstable, and institutions promoting or overseeing research are new[5] and evolving. Complexities like the composite nature of interventions add a further challenge to the ethical management of uncertainty. Futures are also highly uncertain: will cystic fibrosis be cured, or simply transformed into a chronic disease? Will technologies used to control unimaginable suffering in persons with rare genetic diseases lead to applications that enable the comfortable to boost their performance capacities?

Uncertainty, to paraphrase sociologist Ulrich Beck,[6] is the companion of opportunity. Small, well-designed studies can result in significant advances, informing lines of research not contemplated by investigators. For treatment-refractory volunteers who enter gene-transfer studies, uncertainty is far preferable to the inflexible certainty of their prognosis.[†] And for members of the public and policy-makers, uncertainty provides occasion to imagine – and work towards – better and healthier lives.

But uncertainty has a dark side as well. Desire insinuates itself into calculation and measure, and extreme views flourish: perhaps because they are unfalsifiable by conventional means, or perhaps because they provide refuge from the anxiety of uncertainty. Thus, exuberant predictions or fearless experimental ventures at best go unchallenged, and at worst, are welcomed without qualification. Both nightmarish and fanciful projections from critics pass for serious debate, sometimes obscuring more pressing, immediate concerns.

Housekeeping functions – careful record keeping for long-term follow-up, or standard setting for vector dosage, for example – are subordinated to activities that, on their face, seem to promise immediate relief from uncertainty. We see patterns, and base our actions on them, where none exist. Therapeutic interpretations of first-in-human studies are seen as inspirational – something that needs to be protected to preserve the volunteer's sense of hope. Equally troubling is the frequent, strategic mobilization of uncertainty, as when individuals use uncertainty to justify human experiments ("until we go into humans, we'll never know whether the thing works") and then certainty to assure their safety ("well, we've done it in a dozen monkeys and scores of rats, and never had a problem").

[†] Behavioral economists call this "ambiguity preference:" when facing high probabilities of losses (e.g. terminal illness), people tend to prefer decisions that involve uncertainty over those involving known highly probable adverse outcomes.[7]

When people are uncertain, their actions are often informed by subtle cues that slip beneath awareness. Though not documented empirically in this book, there is a strong case to be made that such factors as the metaphors used to describe an intervention, the attire and gestures of clinical investigators, language and terminology, the visible desperate patient (Anderson's "cancer card," but also its antithesis, the injured research subject) all configure the environment in which decisions are made about a study's merits, risks, and alternatives. Experience tells us that this can sometimes lead to suboptimal decision-making.

This book has attempted to identify major sources of uncertainty, and to recommend ways of managing and reducing its untoward consequences. I have thus offered a framework for deciding subject selection, and a heuristic for characterizing value that could inform the design and ethical review of research protocols. I have offered an approach for deciding when to initiate first-in-human trials, and a set of considerations that policy-makers, ethicists, and researchers might use to decide whether and what type of first-in-human trial is warranted. I have sought greater clarity on the ethics of performing translational trials in disadvantaged communities, and offered practical ways that these studies can be rendered consistent with international consensus on human research ethics. Last, I have explored the way investigators and others shape patient and public expectations, and suggested the need for a more robust ethical analysis of this realm.

Gene transfer as a window into the management of uncertainty

With respect to the second thesis, the ethical issues confronting Perin's attempt to translate cardiology cell transfer recapitulate many encountered by gene transfer years earlier. To be sure, there are major differences: whereas somatic (and "therapeutic") gene transfer is today largely uncontroversial, research involving the former takes place amid fractious and persistent debates about human embryonic stem (hES) cells. One consequence is that successful cell-transfer trials are immediately heralded by opponents of hES research, often to the chagrin of the research teams. Medical care and research are also significantly more global today than in the early days of gene transfer. For example, whereas the USA dominated the first decade of gene-transfer research, Europe was the first out of the gate for many cardiology cell-transfer studies. And in Perin's study, we see significant contributions from researchers outside the traditional pharmaceutical axis of North America, Europe, and Japan. As a consequence, discussions about regenerative medicine protocols occasionally take on themes of national competitiveness.[‡]

Nevertheless, parallels abound. First, of course, is the scientific uncertainty: though there is extensive experience of using cells derived from bone marrow for treating hematological disorders, Perin's was one of the first to inject them into heart tissues. The mechanism by which stem cells restore heart function (assuming they do) continues to elude the researchers. Then there are the safety uncertainties. In a related cardiology protocol involving cells derived from muscle (myoblasts), several volunteers developed arrhythmias. This took some researchers by surprise, as similar events had never been observed in over a decade of preclinical and animal testing.[8]

Consider the technical complexity. Like gene-transfer vectors, bone marrow cell preparations involve a mixture of different cell types. Cell interventions often defy straightforward

‡ "We will be licked by the Germans or the Japanese if we don't keep our pace going," says Vincent Pompili, who directs interventional cardiology at Case Western Reserve University in Cleveland, Ohio.[8]

standardization: their properties can change depending on how they are handled outside the body, or how exactly they are injected. Factors like age and comorbidities might influence the properties of stem cells, in which case outcomes might not be comparable across different patients or doses. Another issue is the technique used to deliver agents to the heart muscle. Perin, for example, modified a device (a NOGA® catheter) to deliver cells to the heart; this further complicates assessment and evaluation of safety. And in future protocols, other components – for example, angiogenesis factors that promote cell engraftment – might be necessary as well.[9]

The actors are similar as well. Texas Heart Institute is a hotbed of bold and ambitious clinical champions of the sort who pioneered gene transfer. Since its founding by Denton Cooley[§] in 1962, the center has been among the first to transplant human hearts, to test artificial and xenobiotic hearts, and to implant the totally self-contained mechanical heart, Abiocor.[11] Surveying the rapid uptake of stem cell transfer in interventional cardiology studies, one researcher observed "cardiologists are just animals – extremely aggressive, inquisitive, [with] no rules … They want to do everything yesterday and not today."[8] Many investigators in this field are also entrepreneurs, holding intellectual property and sitting on advisory boards of new biotechnology companies.[¶]

Like gene transfer, cell transfer has been riven with extended debates about the wisdom of initiating trials.[13–15] On one hand, cardiologists like Perin insist that there is no alternative but to venture into patients: "[Basic scientists] don't have patients dying; I do. Daily I have to look them in the eye."[16] On the other hand, researchers like Stanford's Irving Weissmann consider launching studies folly: "these studies are premature and may in fact place a group of sick patients at risk."[17] Such concerns mingle with debates about the purpose and value of first-in-human studies. One researcher heavily involved in preclinical – and as of this writing, clinical-studies warned in 2005 "we are spending a lot of time on clinical trials that are unlikely to give us definitive answers because the science has not caught up yet."[17]

There are similar questions of subject selection in the two fields, with some favoring stable subjects and others preferring the treatment-refractory.[14] Perin's protocol opted for the latter. Inevitably, he conceived the study as a therapeutic opportunity for his patients: "It's ludicrous to say we must understand the molecular mechanisms before we can try anything."[17] His study did, at the outset, have therapeutic design qualities: a single theoretically active dose was delivered, and cells were – by definition – customized for each patient. Yet his study also enrolled controls, and subsequent ones have included sham controls, which involve quite a bit more burden (volunteers have their bone marrow drawn and their hearts catheterized without receiving intervention; until the study is completed, no volunteers are told whether they received active or inert materials).

I have already alluded to the Texas Heart Institute's elaborate choreography of expectation management. With the click of the mouse, anyone can view a series of videos on adult stem cells, in which expectant patients describe the opportunity to participate in the trial as "a miracle," and investigators gently intone the promise of adult stem cells for mending broken hearts.[18]

---

§ Cooley is said to have been asked during a trial whether he was the best heart surgeon in the world. "Yes" he replied, to which the cross-examiner asked "Don't you think that's being rather immodest?" Cooley responded "But remember I'm under oath."[10]

¶ For example Emerson Perin sits on the cardiovascular advisory board of Cytori, a company that is developing adipose-derived cells for the treatment of cardiovascular diseases.[12]

Policy and regulation for cell transplantation, as for gene transfer, are also unsettled. Regulatory agencies find themselves having difficulty navigating criteria for approval, as when the FDA reversed advisory committee recommendations when it denied licensure for a cell-transplantation technology.[19] (For a vigorous defense of the FDA's decision, see DeVita[20].) Because of the heterogeneity of products, regulatory agencies also confront challenges establishing standards and guidelines, and then harmonizing these across jurisdictions. Some also warn of emergent knowledge and management gaps between drug regulatory bodies and research agencies.[21]

Finally, just as several key gene-transfer studies have recruited volunteers in LMICs, Perin pursued his trial in Brazil.** In this particular case, there are differences that might matter from a moral perspective. Rather than importing volunteers, Perin exported the protocol and extensively involved Brazilian medical personnel. This provided training opportunities; but it may have made demands on the local health system. And in this instance, the investigators nod toward the principle of responsiveness: as one of the study's co-authors stated in press materials "If stem cell injection is shown to work, it would be a relatively inexpensive therapy. Also, there would be no need for immunosuppressive drugs to prevent rejection, because patients are injected with their own cells ... this therapy would be generally applicable around the world."[4]

Cell transplantation is but one example of a platform in which ethical challenges in translation recapitulate those for gene transfer. There are many others. Targeted cancer therapies, for example, consist of small-molecule drugs or large molecules like monoclonal antibodies. What they have in common is that they are "rationally designed" (meaning that they are developed from biological principles rather than identified from mass screens). They often take aim at new biological pathways and are highly specific. As noted in earlier chapters, the degree of specificity, in principle, should improve the safety of these drugs by reducing off-target effects. At the same time, however, it also makes prediction of safety from animal models all the more dubious. Targeted cancer therapies also present challenges to translational trial design. Because chemotherapies operate by poisoning tumors, phase 1 trials traditionally are designed to determine the highest possible dose that can be administered without killing patients. With targeted therapies, however, the relationship between dose and effect is less straightforward. Instead, phase 1 trials need to collect information on biological effects like receptor binding. Issues like these present new challenges and introduce layers of uncertainty for translational researchers.[23,24]

Xenotransplantation involves the use of animal tissues or organs in human beings. The uncertainties associated with this technology are perhaps extreme versions of the kind described in this book. The major concern here is the possibility that new animal pathogens will be conveyed to human populations through xenotransplantation protocols.[25] Such risks are extremely difficult to quantify in advance, and depend heavily on human behaviors like sexual practices and hygiene. Given the incalculable odds and high stakes, the International Xenotransplantation Association has taken pains to elaborate ethical principles and practices for first-in-human studies,[26] and censured research teams that transgress their policy.[27]

A number of new brain interventions pose significant ethical challenges at the interface between the laboratory and clinic. One example is trophic factors – proteins that in laboratories protect neurons or stimulate their growth – delivered to the brain to slow or reverse neurodegeneration. Human capacities like language and personality are virtually impossible to effectively model in animals. Because brain function depends on neural connections that

** Other examples of trasnational cell-transplantation studies in cardiology include Patel *et al.*[22]

emerge through development, iatrogenic derangements can be irreversible. That CNS interventions have the highest failure rate during drug development is testimony to the stubborn uncertainty of this research area.

Nanotechnology exploits the unusual mechanical, electrical, chemical, or optical properties of matter at the atomic level. A number of medical applications of nanotechnologies are being pursued, including nanoparticles that "stain" the perimeter of tumors so that surgeons can spot errant metastatic tissue, nanospheres that help drugs across the blood–brain barrier, and structures that provide scaffolding for regenerating damaged tissue.[28] As one might expect with a technology that exploits forces previously inaccessible to human ingenuity, nanotechnologies present a number of major challenges to policy-makers, many of which are like those faced by gene transfer. An FDA task force summarized some of these challenges as follows:

> There may be a fundamental difference in the kind of uncertainty associated with nanoscale materials compared to conventional chemicals, both with respect to knowledge about them and the way that testing is performed ... several recent scientific reviews conclude that the state of knowledge for biological interactions of nanoscale materials is generally in need of improvement to enhance risk assessments and better support risk management decisions.[29]

For instance, nanomaterials come in many shapes, sizes, and purities: how should engineers and regulators characterize their composition?[30] Nanomaterials will likely have many unexpected toxicities. For instance, many nanomaterials are capable of being absorbed through the skin, crossing the blood–brain barrier or being carried along the length of nerves, and they often provoke strong immune reactions. How will investigators monitor toxicities, and how will they decide when to initiate human testing? Nanomaterials might have important public health and ecological impacts, because they persist in the environment;[31] what duties do researchers, sponsors, and host institutions owe to research bystanders, and will new oversight mechanisms – akin to those of the RAC – be necessary to plug the void separating environmental protection and drug regulation? Nanotechnologies are flourishing at the interface of the academy and the private sector. Similar cycles of hope and dread characterize public discourse surrounding them. How will researchers negotiate institutional structures and public engagement?

Finally, these disparate lines of research converge in ways that promise yet greater combinatorial complexity and uncertainty: gene-transfer approaches that employ nanotechnology to enhance delivery; "embryonic" stem cells made by using gene transfer to de-differentiate adult tissues; targeted cancer therapies employed in conjunction with immunotherapy.

Ethical frameworks attuned to uncertainty

The third thesis concerns the problems of extending concepts developed around large-scale, randomized controlled studies to first-in-human studies of novel interventions. Ethics and regulation of research risk in later-stage trials have tended to concentrate on technical procedures. But as we have seen, novel fields draw ambitious and mercurial clinical investigators who create a different kind of risk. Effective risk-management strategies require that these social and institutional factors be addressed. They also require that we look beyond individual studies to structures that enable a field to identify, quantify, and reduce risk. The main recommendations of the book are summarized in Table 10.1.

A concept of value that looks to how medical practice will be modified offers little to the appraisal of studies like Perin's bone-marrow study. Instead, I have defended a translational

Table 10.1 The ethical management and response to uncertainty in translational clinical trials involving novel agents

(1)	Consider social sources of risk
	e.g. tournament incentive structure, clinical champions, unstable rules of conduct
(2)	Devise ways of pooling risk information within and across studies
	e.g. standardizing reagents and trial designs, sharing toxicology data
(3)	Risks in translational trials should be ethically justified by knowledge benefits
	e.g. exclude direct benefits (e.g. therapeutic value for subjects) and collateral benefits (e.g. extra monitoring) from risk–benefit analysis
(4)	When uncertainty is great, assume that protocol is higher risk
	e.g. use "upper-bound" rather than "best-guess" estimate of risk
(5)	Prospective evaluation of knowledge benefits should be based on expectation that both "positive" and "negative" findings will be meaningful and informative
	e.g. design hypothesis testing like correlative studies into translational trials
(6)	Maintain modest translational distance between preclinical and clinical experiment
	e.g. minimize number of assumptions linking preclinical and clinical studies by maximizing: internal and external validity of preclinical studies, correspondence with circumstances of clinical study, and credibility of supporting data
(7)	When pursuing trials in disadvantaged populations, ensure responsiveness
	e.g. sponsors/research teams should show evidence that study addresses urgent health need of host community and that results can be deployed in that setting – either directly or indirectly through philanthropic intervention or intellectual property agreements
(8)	Manage public expectations
	e.g. foster transparency; leadership should "prime" and "dampen" expectation as necessary; check content and gestures that present trials as therapeutic ventures.

model that presses first-in-human trials to anticipate how "negative results" will inform laboratory researchers, and to design studies with iterative and reciprocal considerations in mind.

For over a decade, equipoise has provided ethics committees and investigators with a basis for determining whether and when it is appropriate to randomize volunteers to different treatments. The uncertain knowledge environment and statistical underpowering make it difficult to use equipoise as a basis for deciding if a first-in-human trial should be initiated, and if so, whether to pursue exploratory testing, dose-escalation design, placebo controls, etc. Instead, I have offered the heuristic of "translational distance."

Related to this, the well-established practice of component analysis in the ethical analysis of trial risk confronts many challenges when applied to first-in-human studies. Conventionally, interventions have been categorized as "therapeutic" such that their risks are justified by appeal to therapeutic benefit. Clearly, intention is insufficient to guide decisions concerning whether or not an action is classified as "therapeutic" or "research:" a medical procedure should not count as therapeutic if one can foresee unintended and undesirable consequences that will occur as a result. I have thus argued that first-in-human trial risks are justified only by appeal to societal benefits.

With respect to international studies, ethical analysis and debate have tended to center on questions of post-trial access. Yet for studies like Perin's and those examined in this book, post-trial access has little to offer: phase 1 studies are not designed to vindicate interventions, and strategies like Perin's are aimed at durable effects. Some ethicists argue that something equivalent to post-trial access can be achieved by offering benefits, such as training opportunities and equipment, to host communities. But how is this approach applied when study volunteers are transported out of the local context to medical centers in high-income countries? I have instead concentrated on "responsiveness" as the condition that anchors recruitment of deprived populations into translational clinical trials.

Expectation management and communications with various publics have not been focal points for ethical discussion and policy. There are no doubt instances of expectation management debacles in later-stage studies. Nevertheless, such issues take on greater significance in early-stage studies, where the therapeutic properties of interventions are supported by powerful conjecture but total absence of clinical evidence, where information access about trials is highly restricted, and where the consequences of mismanaged expectations around one trial can tarnish the reputation of investigators pursuing related but distinct research projects. I have attempted to place expectation management center stage in the ethics, and offer some initial steps towards epistemologically satisfying policy on public communications of clinical researchers.

Agendas for further research

This book is hardly the final word on this subject matter, and there are a number of ways the analyses offered in these pages might be extended. I divide these into four different research agendas that, I hope, this volume might inspire.

An agenda for this book's proposals

A general problem for research ethics is that policies and frameworks offered typically fail to describe criteria for measuring their success. It's worth thinking ahead, then, to how the frameworks, policies, and practices advocated here might be tested in a way that leads to their rejection, embrace, or refinement. I suggest four criteria against which my various prescriptions should me measured.

Internal consistency: Do my various policy prescriptions tug in opposing ways on a given issue, or openly contradict each other? For example, does my proposal of "translational value" lead to an escalation of burden and risk that undermines the goals of enhancing patient welfare and building confidence in the translational research enterprise?

Coherence: Do the arguments presented in this text cohere with other policies and practices that have withstood tests of critical analysis and implementation? For example, environmental protection, worker safety, pension management, and catastrophe insurance all deal with measuring and managing uncertainty. There are many ways that each of these different areas can bring insights to one another, and many ways in which the thinking, underlying policies, and practices in these different areas contradict one another.

Feasibility: Can the proposed policies be implemented in the diverse settings where translational research occurs, can they be extended from gene transfer to other novel technologies, and will implementing recommendations require substantial investment in human and/or material resources? Does their implementation sacrifice or undermine otherwise important values or policy goals?

Consequences: are the consequences of proposals offered here, on balance, favorable – and more favorable than alternative proposals? For example, would my concept of translational value actually inform how investigators design trials and how IRBs review them? And if so, is there any evidence that it would actually hasten the development of effective medical interventions while reducing the expenditure of resources?

An agenda for uncertainty and ethics

The next research agenda concerns systematic inquiry into the problem of uncertainty, risk, and knowledge in scientific medicine. How do study volunteers make decisions when outcomes are highly uncertain? What kinds of bias influence the interpretation by clinical investigators of small-scale studies? Do clinical investigators prepare adequately for rare, high-consequence events? To what extent do deliberative procedures like independent review – and centralized review – screen out decisions that are based primarily on affective risk perceptions? What kinds of cognitive heuristics do policy-makers and ethics reviewers use when assessing risks and societal benefits, especially when both are abstract and indeterminate? And do these distort decision-making?

The literature on uncertainty, risk, and behavioral economics is burgeoning. With a few exceptions,[32] it has yet to be applied to human research ethics and in particular, the types of trial that are characterized by the profoundest types of uncertainty. Psychologists Daniel Kahneman and Amos Tversky have described numerous ways that biases creep into decision-making when individuals are confronted with uncertainty. For example, individuals tend to discount or ignore evidence when it contradicts prior beliefs, and they tend to trust evidence when it supports prior beliefs ("confirmation bias"). I have already suggested that there are a number of reasons – denial processes for patients, personal investment for translational researchers, for example – that accentuate tendencies toward confirmation bias in translational trials. Another important discovery from this area is the "law of small numbers:" the tendency to make excessively strong inferences about the existence of patterns after a small number of observations. This has been documented even in situations where observers have a high degree of statistical sophistication.[33] Another bias that might shed light on how various parties to gene-transfer studies make decisions about trial initiation and risk is "action bias:" a tendency to favor action over inaction. On its face, action bias would seem to contradict the well-established "status quo bias," in which decision-makers prefer inaction. However, according to some researchers, circumstances like uncertainty about the consequences of action, ease with which positive outcomes of actions can be imagined, and the ability for an individual to take credit for the outcomes of an action will tend to tilt decision-making toward action.[34]

Is an atmosphere of risk and uncertainty also protective in certain ways? There is some evidence that automobile drivers compensate for perceived safety – because their cars are equipped with airbags, or because they drive a large vehicle – by taking greater risks on the road.[35,36] Though this "risk homeostasis" model is disputed,[37,38] the suggestion that persons proceed more cautiously when they perceive danger is plausible and has implications for risk management in translational research. One question is how policy-makers should manage the safety expectations of researchers or institutions charged with managing safety.

Observations like these call out for empirical and theoretical investigations of how findings from early studies are interpreted and applied, the mechanisms and institutions we currently use to evaluate the merit and risks of research proposals, and the ways that investigators and others manage research risks.

An agenda for research ethics

Among the many possible problems for research ethics raised or grazed by this book, five stand out for warranting further work. The first is the problem of translational distance. As described in Chapter 7, there are a number of aspects of this proposal that need refinement: what is an appropriate "benchmark" for translational distance, for example? How might this be operationalized?

A second problem raised by this book concerns the research team's obligations to various publics other than the study volunteer. Chapter 9 attempted to put expectation management on the agenda of research ethics. Much will need to be worked out. For instance, at a micro level, what responsibilities do investigators have in managing the expectations of research volunteers, and/or their families? At a macro level, what role should research teams play in the expectation management of members of the public? What are the appropriate "rules" of expectation management, who should police them, and how?

The third concerns agenda setting in translational research. As noted in Chapter 2, when gene transfer was first proposed, it was viewed largely in terms of addressing rare, genetic disorders. Since then, cancer and cardiovascular disease have overtaken genetic disease as the prime targets for gene transfer. How should emergent fields set – and maintain – their research agenda? Obviously, translational researchers merely respond to incentives and policies that reward research in some areas and penalize research in other areas. But researchers arguably have a responsibility to help shape the research agenda, to tutor publics on research areas of pressing importance, and to "nudge" the scientific research enterprise toward addressing major social priorities.[39] To what extent are researchers adequately directing their energies toward social priorities? And what are the appropriate means of "nudging" a research enterprise?

The fourth issue concerns transgenerational research ethics. How will inheritable gene-transfer applications be evaluated for safety and efficacy? Existing regulatory and review systems are designed to protect existent research subjects, rather than incipient or future ones. Any attempt to extend gene transfer to the germline will require extensive ethical and policy work in this area.[40,41]

The last issue was only touched on briefly in this book: centralized and transparent review of protocols. Gene transfer, along with a special category of pediatric trials (so called "407" studies),[42] are unique in that they are subject to a procedure for centralized ethical review. I believe strongly – but have not made the case in this book – that such a forum provides a critical mechanism for regulating and managing risks, uncertainties, and public expectations in novel research areas. But can this be shown empirically? For instance, one counter-argument against centralized review is that gene-transfer researchers encounter a number of risk and expectation problems despite the existence of centralized review. Do policy-makers, investigators, and IRBs actually absorb the recommendations of centralized review structures? For instance, the Parkinson's disease study with which Chapter 7 opened was clearly approved by Cornell's IRB despite its highly skeptical reception at the RAC. Do centralized review structures result in moral hazard, whereby investigators (and IRBs) punt their moral responsibility to central, expert review agencies? A similar concern has been raised in the context of increasingly stringent IRB review.[43] Do centralized review structures attend to the moral richness of the ethical problems encountered in studies? Are they unduly swayed by the testimony of research advocates, for example? Why is it that, in almost two decades of operation, serious discussion of basic justice issues in gene-transfer research – like those I describe in Chapter 8 – have never been raised? Does centralized review showcase an impoverished, "book-keeping" approach to ethical review of new technologies, as some have suggested?[44]

An agenda for gene-transfer ethics

The primary focus of this book has been the ethical challenges encountered in testing gene-transfer interventions for the first time in human beings. This hardly exhausts the range of policy and ethical challenges.

The first two issues have been debated since gene transfer was first proposed. However, whereas initial discussions were largely speculative and abstract, policy-makers are at present confronting gene-transfer applications that give these debates a salience they once lacked. The first is germ cell gene-transfer: under what circumstances is it ethical to intentionally modify human germ cells? With techniques that transfer "healthy" mitochondria into defective ova ("ooplasmic transfer"), infertility specialists have arguably already "crossed the line" into inheritable genetic modifications.

In response to initial forays into ooplasmic transfer in the late 1990s, the US FDA declared that such procedures would be regulated as investigational new drug applications.[45] This policy response was largely driven by concerns about the safety of the procedure. But are there defensible grounds for regulating ooplasmic transfer – or more direct forms of germ cell modification – if risks can be reduced to an acceptable level? What about other forms of inheritable interventions, like in-utero gene transfer? I suspect that social and ethical consensus on this issue will be as elusive as it is for other debates involving human embryos.

Closely related to this is the ethics of genetic enhancement (that is, the application of gene transfer for traits that improve human capacities or create new ones, rather than restoring ones that aren't functioning properly). Are there compelling ethical and/or policy grounds to limit or ban certain types of somatic and/or germ cell genetic enhancement technologies? As noted in Chapter 2, a number of gene-transfer applications aimed at "lifestyle" have already been tested; other interventions are aimed at life-threatening disorders, but have obvious cosmetic applications (e.g. gene-transfer strategies aimed at treating muscle wasting might also be used to build muscle mass in athletes). Again, I believe that there is a compelling – but hardly airtight – case to be made for regulating such technologies, but I suspect that widespread social and scholarly consensus will be elusive.

Making predictions about cutting-edge technologies is fraught with hazard, and I would be foolish not to pay heed to the many unsuccessful predictions for gene transfer. Given promising results in gene-transfer applications involving ADA-SCID, the perseverance and progress of researchers using immunosuppressive regimes to support gene transfer for monogenic disorders like hemophilia, and the fact that phase 3 trial volume has undergone a steep increase over the last several years, it may not be too early to consider a few ethical issues that will require policy responses once gene transfer has been translated to clinical practice.

The first concerns regulation. The X-SCID study with which I opened Chapter 4 provides an instance of a life-saving intervention that is nevertheless associated with a significant risk of life-threatening adverse events. What is the appropriate regulatory stance on such "heroic" interventions? What are the appropriate procedures for making such policies? Related to this, in Chapter 4 I described how many gene-transfer interventions, because they involve durable exposure to transgenes, might involve adverse events that emerge after a latency period. Should drug regulators establish pharmacosurveillance programs that are custom tailored for intervention classes like gene transfer? If so, what are the most effective ways of ensuring that these detect rare, latent toxicities?

A second issue concerns cost and access. Though gene transfer is often presented as a low-cost alternative to chronic drug treatment regimes, new medical technologies, according to economic analysts, account for 10–50 percent of the increase in healthcare costs in the USA after World War II.[46,47] There are a number of characteristics of gene-transfer interventions that threaten to undermine their cost-saving potential. Many target orphan diseases. Because of policies in many countries that are designed to reward companies developing drugs for rare diseases, orphan drugs tend to be extremely expensive. One-time interventions also concentrate expenditures at a single time, rather than spreading costs across the course of an illness. This tends to deter health insurers from providing coverage, because they have no assurance that they will recover cost savings. Because of their complexity and risks, good manufacturing practice is likely to be exacting and costly. Whereas with many drugs, policy-makers wishing to contain healthcare costs can look to generic products, the pathway toward and effect of licensure for generic biologics – especially complicated ones like gene-transfer products – are very unclear.[48, 49]

Uncertain conclusions

It is often said that technological innovation is outstripping society's capacity to develop and implement sound social policies. I have always been skeptical of this claim, suggesting as it does a deterministic relationship between technology and people, and underestimating the creative capacities of human beings and their social groups to make sense of technologies and respond to them.

Gene transfer is a case in point. Now, eighteen years since the first "official" human trial was initiated, clinical applications are only beginning to emerge from the haze of uncertainty. And though they promise relief from suffering for many, like the drugs that preceded them, gene-transfer products are likely to be imperfect – ambiguously effective in some cases, dangerous in others, and costly in still others.

This modest but steady pace of clinical translation has afforded policy-makers, ethicists, scholars, and scientists ample time to reflect on policy and practice around translational research and gene transfer, and to anticipate the challenges that confront researchers like Emerson Perin, who want to carry medical knowledge into new realms.

Uncertainty will be intrinsic to their endeavor. As long as clinical investigators venture toward new, unfamiliar research areas, they and their research subjects will necessarily confront risks that are incalculable. It is tempting to view their encounter with risk, uncertainty, and mishap through the eyes of Kafka's Hunter Gracchus, who wanders at sea aimlessly, eternally, and resignedly. I think we can do better. There are many ways that we – ethicists, IRB members, funders, policy-makers, investigators, referring physicians, scientists, disease advocates, university administrators, investors, biotechnology firms, patients, research subjects, spectators – can pause, learn from experience, and "adjust our eyes" to the "darkness."

References

1. Orlic D, Kajstura J, Chimenti S, Jakoniuk I, Anderson SM, Li B, *et al.* Bone marrow cells regenerate infarcted myocardium. *Nature* 2001; **410**(6829): 701–5.
2. Perin EC, Dohmann HF, Borojevic R, Silva SA, Sousa AL, Mesquita CT, *et al.* Transendocardial, autologous bone marrow cell transplantation for severe, chronic ischemic heart failure. *Circulation* 2003; **107**(18): 2294–302.
3. Wendler R. Will stem cells restore life to failing hearts? *Texas Medical Center News* 2004 April.
4. Media Contact – Merville S. *Press Release: Willerson leads stem cell clinical trial to treat heart failure – Houston project follows encouraging results in Brazil.* Houston: UT Health Science Center, 2004 March 26.
5. Aguilar LK, Aguilar-Cordova E. Evolution of a gene therapy clinical trial. From bench to bedside and back. *J Neurooncol* 2003; **65**(3): 307–15.
6. Beck U, Willms J, Pollack M. *Conversations with Ulrich Beck.* Cambridge,UK, Polity Press, 2004.
7. Camerer C, Weber M. Recent developments in modeling preferences: Uncertainty and ambiguity. *J Risk Uncertainty* 1992; **5**(4): 325–70.
8. Couzin J, Vogel G. Cell therapy. Renovating the heart. *Science* 2004; **304**(5668): 192–4.
9. US Food, and Drug Administration. Biological Response Modifiers Advisory

Committee Meeting, March 18, 2004. US Department of Health and Human Services, 2004; 106–7.

10. Altman LK. The feud. *The New York Times* 2007 November 27.
11. SoRelle R. Third abiocor artificial heart implanted in Houston. *Circulation* 2001; **104**(15): E9033–4.
12. "Cytori Therapeutics appoints Cardiovascular Clinical Advisory Board" March 21, 2006. Available at: http://ir.cytoritx.com/releasedetail.cfm?ReleaseID=191767. Last accessed June 4, 2009.
13. Bauersachs J, Thum T, Frantz S, Ertl G. Cardiac regeneration by progenitor cells – bedside before bench? *Eur J Clin Invest* 2005; **35**(7): 417–20.
14. Rosen MR. Are stem cells drugs? The regulation of stem cell research and development. *Circulation* 2006; **114**(18): 1992–2000.
15. Prockop DJ, Olson SD. Clinical trials with adult stem/progenitor cells for tissue repair: let's not overlook some essential precautions. *Blood* 2007; **109**(8): 3147–51.
16. Fox C. Can stem cells save dying hearts? Just ask this pig. *Discover Magazine Online* 2005 September 9.
17. Wade N. Tracking the uncertain science of growing heart cells. *The New York Times* 2005 March 14.
18. See, for example, "Innovations in Stem Cell Research" and "Rebuilding the Heart" at: www.texasheart.com/AboutUs/News/media_room.cfm (last accessed July 21, 2008).
19. The regulator disapproves. *Nat Biotechnol* 2008; **26**(1): 1.
20. DeVita VT Jr. The Provenge decision. *Nat Clin Pract Oncol* 2007; **4**(7): 381.
21. Rosen MR. Are stem cells drugs? The regulation of stem cell research and development. *Circulation* 2006; **114**(18): 1992–2000; at 1997.
22. Patel AN, Geffner L, Vina RF, Saslavsky J, Urschel HC Jr, Kormos R, *et al.* Surgical treatment for congestive heart failure with autologous adult stem cell transplantation: a prospective randomized study. *J Thorac Cardiovasc Surg* 2005; **130**(6): 1631–8.
23. Parulekar WR, Eisenhauer EA. Phase I trial design for solid tumor studies of targeted, non-cytotoxic agents: theory and practice. *J Natl Cancer Inst* 2004; **96**(13): 990–7.
24. Booth CM, Calvert AH, Giaccone G, Lobbezoo MW, Seymour LK, Eisenhauer EA. Endpoints and other considerations in phase I studies of targeted anticancer therapy: recommendations from the task force on Methodology for the Development of Innovative Cancer Therapies (MDICT). *Eur J Cancer* 2008; **44**(1): 19–24.
25. Weiss RA. Xenotransplantation. *BMJ* 1998; **317**(7163): 931–4.
26. Sykes M, d'Apice A, Sandrin M. Position paper of the Ethics Committee of the International Xenotransplantation Association. *Xenotransplantation* 2003; **10**(3): 194–203.
27. Sykes M, Cozzi E. Xenotransplantation of pig islets into Mexican children: were the fundamental ethical requirements to proceed with such a study really met? *Eur J Endocrinol* 2006; **154**(6): 921–2; author reply 923.
28. Buxton DB, Lee SC, Wickline SA, Ferrari M. Recommendations of the National Heart, Lung, and Blood Institute Nanotechnology Working Group. *Circulation* 2003; **108**(22): 2737–42.
29. US Food, and Drug Administration. *Nanotechnology: A Report of the US Food and Drug Administration Nanotechnology Task Force.* US Department of Health and Human Services, 2007 July 25.
30. Holsapple MP, Farland WH, Landry TD, Monteiro-Riviere NA, Carter JM, Walker NJ, *et al.* Research strategies for safety evaluation of nanomaterials, part II: toxicological and safety evaluation of nanomaterials, current challenges and data needs. *Toxicol Sci* 2005; **88**(1): 12–17.
31. Wilson RF. Nanotechnology: the challenge of regulating known unknowns. *J Law Med Ethics* 2006; **34**(4): 704–13.
32. Weinfurt KP. Value of high-cost cancer care: a behavioral science perspective. *J Clin Oncol* 2007; **25**(2): 223–7.
33. Tversky A, Kahneman D. Belief in the law of small numbers. *Psychol Bull* 1971; **76**(2): 105–10.
34. Patt A, Zeckhauser R. Action bias and environmental decisions. *J Risk Uncertainty* 2000; **21**(1): 45–72.

35. Wilde GJ. Does risk homoeostasis theory have implications for road safety. For. *BMJ* 2002; **324**(7346): 1149–52.

36. Wilde GJ. Risk homeostasis theory: an overview. *Inj Prev* 1998; **4**(2): 89–91.

37. Robertson LS, Pless IB. Does risk homoeostasis theory have implications for road safety. Against. *BMJ* 2002; **324**(7346): 1149–52.

38. O'Neill B, Williams A. Risk homeostasis hypothesis: a rebuttal. *Inj Prev* 1998; **4**(2): 92–3.

39. Kitcher P. *Science, Truth, and Democracy.* New York, Oxford University Press, 2001.

40. Dresser R. Designing babies: human research issues. *IRB* 2004; **26**(5): 1–8.

41. Dresser R. Genetic modification of preimplantation embryos: toward adequate human research policies. *Milbank Q* 2004; **82**(1): 195–214.

42. US Food, and Drug Administration. *Public Welfare: Protection of Human Subjects – Additional Protections for Children Involved as Subjects in Research* 45CFR46.407. US Department of Health and Human Services.

43. Moreno JD. Goodbye to all that. The end of moderate protectionism in human subjects research. *Hastings Cent Rep* 2001; **31**(3): 9–17.

44. Evans JH. *Playing God?: Human Genetic Engineering and the Rationalization of Public Bioethical Debate.* Chicago, University of Chicago Press, 2002.

45. Kimmelman J. The ethics of human gene transfer. *Nat Rev Genet* 2008; **9**(3): 239–44.

46. Weisbrod BA. The health care quadrilemma: an essay on technological change, insurance, quality of care, and cost containment. *J Economic Literature* 1991; **29**: 523–32.

47. Congressional Budget Office. *Technological change and the growth of health care spending.* Washington, DC, Congressional Budget Office, January 2008.

48. Frank RG. Regulation of follow-on biologics. *N Engl J Med* 2007; **357**(9): 841–3.

49. Dudzinski DM, Kesselheim AS. Scientific and legal viability of follow-on protein drugs. *N Engl J Med* 2008; **358**(8): 843–9.

Epilogue

The science, ethics, and policy of gene transfer are rapidly evolving. As a companion to this book, I maintain a running commentary on developments in ethics, gene transfer and cell transplantation, and translational clinical trials at my blog *Lost in Translation* (http://lostintranslationethics.blogspot.com/)

The blog is designed to report on relevant news items, identify emerging ethical issues in translational research, apply frameworks developed in the book, and extend them; readers are invited to use the forum to discuss and debate the ethics of translational clinical research.

Index